# STUDIES IN STARLIGHT

*Understanding Our Universe*

To my wife, Karen, and to her special creative efforts: Stephen Kingsley Caes and Pamela Lynn Caes

# STUDIES IN STARLIGHT

## *Understanding Our Universe*

Charles J. Caes

TAB BOOKS Inc.
Blue Ridge Summit, PA

FIRST EDITION
FIRST PRINTING



Library of Congress Cataloging in Publication Data

Caes, Charles J.
Studies in starlight.

Bibliography: p.
Includes index.
1. Astrophysics. I. Title.
QB461.C34 1988 523.01 87-33515
ISBN 0-8306-0946-6
ISBN 0-8306-2946-7 (pbk.)

Questions regarding the content of this book should be addressed to:

Reader Inquiry Branch
TAB BOOKS Inc.
Blue Ridge Summit, PA 17294-0214

# Contents

# Introduction

THIS TEXT IS DESIGNED TO INTRODUCE YOU TO ASTROPHYSICS by surveying the visible and invisible fields of astronomy. Much of modern astronomy has been turned over to the astrophysicist. This has become necessary because without applying the great wealth of information that has been accumulated by particle physicists to astronomical and cosmological research, the past few decades would never have been the great period of discovery and advancement that they now represent.

Specifically, this book deals with the radiation that is coming towards Earth from outer space, and with the way in which this radiation can be deciphered. This radiation is much more than simply electromagnetic waves; it is one of the languages of the cosmos. It carries with it a great many stories of the past, reports about the present, and prophecies about the future.

The Earth's atmosphere—and sometimes the surface itself—is bombarded with radiation coming at different wavelengths and, therefore, at different frequencies. Some of the radiation is terrestrial in origin, but much of it comes from objects and systems moving well beyond our world. We can see some of the radiation, and that which comes at the visual wavelengths we call *light*. The invisible radiation is that which comes in wavelength and frequency ranges labeled as radio, infrared, ultraviolet, X-ray, and gamma ray. Radar, microwave, and molecular radiation are a part of the radio ranges of the electromagnetic spectrum. This term *electromagnetic spectrum* refers to the entire range of electromagnetic radiation from the radio to the gamma wavelengths.

When we speak of electromagnetic waves, what we are referring to are waves resulting from the acceleration of charged particles. These particles are the very minute bits of matter held by scientists to represent the basic constituents of all that exists. Some of these elementary particles have combined to form that basic unit of all chemical elements called the *atom*.

The new astronomies focus on the electromagnetic radiation coming from outer space. As you treat every alphameric that you read in this text as a graphic message, and as programmers treat electrical pulses as data carrying bits, so, too, astronomers working in the invisible ranges of the electromagnetic spectrum treat each photon of radiation as important signals from nature telling about the past, present, and future.

In order to appreciate the magnificent achievements that astronomers and physicists have made, it is particularly important to know a bit of astronomy's past, men and women's quest for understanding visual light and all other radiation, and the birth, development and progress of the sciences astrophysics and astrochemistry. Thus, this book is developed into two main parts, the first dealing with a history of astronomy and related sciences, and the second dealing with modern research and achievements.

The first three chapters make up Part 1 and deal with history.

Chapter 1 looks at the birth and development of astronomy. As for its birth—well, this must be a subject for speculation for by the time the ancient civilizations began recording their histories, they seem to already have acquired a great knowledge of stellar positions and some planetary movements. As for its history, much has been recorded.

The section on ancient astronomy, covering the period from about 3000 B.C. to 7000 B.C., looks at Egyptian, Sumerian, Babylonian, Chaldean, Persian and Assyrian observations. The section on the late ancient period discusses ancient history in order to fully explain the basis for the ancient Greek separation of science from religion, then proceeds to give a chronology of developments according to the periods in which individual philosopher/scientists lived. There are few books which treat the history of astronomy in this way, and so the chapter becomes a handy reference for a chronological history of astronomy.

The section dealing with astronomy during the middle ages begins with an overview of why and how Greek ideas evaporated for six or seven hundred years and astronomy, for the most part, wound up at a standstill. The section contains an overview of the fall of the Holy Roman Empire and the eventual preoccupation of the Arabians with what was, for them, newly discovered Indian and Greek texts on math, astronomy, and the natural sciences.

In the Early Modern Period, the works of Copernicus, Brahe, Galileo, and Kepler are reviewed, and then the contributions of later geniuses who systematically studied the sky—men like William and John Herschel, Charles Messier, John Dalton, William Cranch Bond, and George Ellery Hale.

When the modern period in astronomy actually began is not easily determined, and different historians establish different dates for its start. For the purposes of this text, it is considered to have begun with the ideas of Herman Mankowski, whose mathematical models of the universe were a break with Newtonian physics and an inspiration for Albert Einstein. Thus, the final section of Chapter 1 begins with a mention of Mankowski's four-dimensional continuum and then proceeds to look at the contributions of Schwarzchild, Hertzsprung, Hess, Russell, Shapley, and others.

Chapter 2 is about the history of studies and experiments with light, beginning with ancient and classical ideas and continuing up to present theories. Light remains a mystery to scientists, even today. It is some strange phenomenon that has the characteristics of both a wave and a particle. Scientists know what to expect of it, how to produce and harness it, but they haven't much of an idea of exactly what it is. But light is all that astronomers see when they look at the heavens. Astronomers are primarily observers of light. Thus, the more they know about it, the more they can learn about the sources from which it appears to come. More startling observations and discoveries about the properties and, perhaps, the nature of light, will come in the future. I would dare to speculate that we hardly know very much about how and why we sense it, of what it actually consists, and what may actually be its true speed, though on the point of its speed there is little doubt in the scientific community.

So Chapter 2 contains a review of the ideas of Pythagoras, Democritus of Abdera, Alhazen, Galileo, Romer, Bradley, Morly, Descartes, Huygens, Planck, Einstein, and many other noted researchers and theorists. And there are descriptions of the pressure, particle, and wave theories of light.

Because of the many subtopics that make up the chapter, two summaries are presented, one at about midpoint in the chapter to summarize the main ideas and facts about light, and a second at the end of the chapter to give an overview of the history and research.

Chapter 3 is a history of the development of astrochemistry and astrophysics. It reviews the contributions of Kirchhoff, Bunsen, von Fraunhofer, Angstrom, Draper, and others, and offers a description of an electromagnetic wave, this time in a much more technical way.

Part 2 begins with Chapter 4, and in this part of the text the chapters are relatively short. Chapter 4 is about visual astronomy and includes sections on the major observatories, types of telescopes, and the important discoveries that continue to be made.

Chapter 5 is concerned with radio astronomy, which deals with radiation from space that may be detected at wavelengths from about 1 mm to 30 m. The chapter looks at the contributions of Karl Jansky and Grote Reber, and then reviews the progress radio astronomy has made since the 1930s. It looks at the solar system, the Galaxy, and other star systems from the perspective of the radio astronomer, and also gives a radio "picture" of supernovae, ionized hydrogen clouds, molecular clouds, pulsars, and maser 5.

Chapter 6 deals specifically with microwave cosmology. Microwaves lie in the radio regions of the electromagnetic spectrum, but they are given a separate chapter because their implications are more cosmological than astronomical. Cosmology is a theoretical science that deals with the universe as a system, whereas astronomy is concerned with the gathering of facts about specific objects within the universe.

Moving along the electromagnetic spectrum from the radio ranges through the visual, you come next to the infrared wavelengths and frequencies. Infrared astronomy is the subject of Chapter 7; this has become a very important branch of astronomy. Contributions to the great storehouse of astronomical data by infrared astronomers has been enormous in recent years. The chapter reviews some of these contributions and then takes a look at the infrared sky.

Chapter 8 moves into the ultraviolet ranges of the electromagnetic spectrum. It begins with the 18th century experiments of Johann Ritter and his discovery of the invisible field on the other side of the violet range in the visual part of the spectrum. Then it looks at the advances in ultraviolet astronomy that have occurred in the past couple of decades. For the most part, researchers in ultraviolet astronomy must get above the Earth's atmosphere to launch their studies and conduct their observations, so they are dependent on space-age technology. What the sky reveals to ultraviolet astronomers once they get their equipment into space is the subject of the rest of the chapter.

Continuing to move further and further away from the radio ranges of the spectrum, you come to the X-rays, and the subject of Chapter 9. X-ray astronomy is a very new science but one that is playing an increasingly important role in astronomical and cosmological research. X-rays themselves are a fairly recent discovery, a discovery dating back to the end of the 19th century when Wilhelm Röntgen was conducting his experiments with cathode-ray tubes.

X-ray astronomers, just like most astronomers working in invisible wavelengths, must get their detecting equipment into space before they can conduct their investigations. But, in addition, they need special types of telescopes, because the refractive index for X-radiation is very different from that for visual light and reflection is not as easily accomplished. After this technical review of problems related to X-ray studies, the remainder of the chapter describes the sky as it is to X-ray astronomers, and what they have discovered or learned.

Chapter 10 looks at gamma ray astronomy and how far it has come since the 1960s. Gamma ray astronomers are also highly dependent on space technology. And, they must, of course, get high above the Earth's atmosphere to operate. Their gamma ray sky is a very barren sky compared to that observed at the other wavelengths, but it is expected that gamma ray detection will play a more important role in astronomy in the next couple of decades.

# Part 1
# Backgrounds

# Chapter 1
# The Art and Science of Astronomy

THE HISTORY OF MAN'S STUDY OF THE HEAVENS can be divided into five distinct periods. The first, the *early ancient* period, covers the earliest known records and the period during which astrology and astronomy were so closely tied that they had no distinction, and astronomy was still mainly a religious or social art. It extends from about 3500 B.C. to 700 B.C. The next period is marked by serious attempts to develop astronomy into a science, though the subject continues to be hampered by religious beliefs and tradition; this is the *late ancient* or *Greco-Roman* period (the latter because the major contributors were the Greeks, whose culture eventually meshed with that of ancient Rome); this period begins sometime in the seventh century B.C. and continues until about 3 A.D. Next is the *medieval* period, extending until about the fifteenth century and marked by a preoccupation with Ptolemic, or classical, astronomy and its verification. Afterward comes a brave attempt to break with the past, even challenging religious doctrine, in order to develop new cosmological concepts and find the basic laws that govern the universe; this is the *pre-modern* period, and it runs through most of the nineteenth century. Finally, there is the *modern* period which, for convenience, begins with the break from the ether hypothesis and the development of a four-dimensional cosmological model. However, one could just as easily argue that the modern period actually began with Einstein's relativity theories, or the detection of radio signals from outer space back in the early '30s.

## ANCIENT ASTRONOMY (3000-700)

The very earliest astronomy was limited to naked eye observations of the positions and movements of things in the sky. First incentives to make positional and motional notations of heavenly bodies were probably inspired by a need for establishing time frames. Archaeological digs and historical researchers have discovered astronomical writings that can be traced to the very earliest civilizations in Babylonia, China, and India, and these indicate a preoccupation with lunar phases and other cycles which would lead to the development of a dependable calendar. At first, astronomy was a very prac-

**Table 1-1. Ancient History of Astronomy (3000 B.C.-700 B.C.).**

| Culture | Comment |
|---|---|
| Egyptian | Calendars since at least 3000 B.C.; 12 months of 30 days; little interest in developing astronomy as a science; kept time by sundials, water clocks and dehanal system; cosmology derived from mythology; before creation, there was an eternal ocean. |
| Mesopotamian | Developed an early fascination for astrology; studied astronomy not only for practical reasons but also to acquire knowledge for its own sake; everything in the universe originated with water. |
| Sumerian | Calendar similar to that of the Egyptians; may have influenced Egyptian science. |
| Babylonian | Close ties with Sumerian and Chaldean cultures; astronomy and astrology definitely rooted in Sumerian ideas; preoccupied with developing a precise calendar; concerned with lunar and solar phases; used a base-60 counting system which we still use to keep time; strongly influenced by Chaldean astrology. |
| Chaldean | Name synonymous with astrology; astronomy only important as basis for astrological soothsaying; respected by other cultures for knowledge of the skies and earth sciences. |
| Assyrian | Semitic origins; copied the cultures of newly conquered people or surrounding cultures; completely absorbed with Chaldean astrology; sought supernatural celestial signals for each personal, political, and military undertaking. |
| Persian | Magi astrology and science, possibly evolved from the Chaldean art; expected the sky to herald the birth of the first of three messiahs that would lead all Zoroastrians to an elevated worldly state; foresook astronomy when they became obsessed with astrology; the universe owes its existence to an eternal deity. |
| Chinese | Obsessed with finding ways to measure the passage of time, their main interest in astronomy until astrology became their obsession; stars foretell love, war, business, and all undertakings. |

tical art. But there were also religious and political reasons for understanding the heavens, as well as the necessity of having a calendar to help government and agricultural planning. To the guardians of the ancient religions there was a divine scheme to heavenly movements that would help man communicate with the gods—or at least understand what the gods wanted. To the ambitious with political designs, a knowledge of astrology was a key to power and fortune.

Egyptian astronomers had fairly accurate calendars as far back as 3000 B.C. (though, when compared to the calendars of today, theirs might seem to mark time on some distant planet rather than here on Earth). The Egyptians had a year with 365 days and 12 months, yet each of their months was assigned 30 days. That, of course, presented them with an unusual dilemma—extra days to spend or carry over to the next year, which would throw their calendar so far off that before long the seasons would reverse themselves. Knowing they had to do something with the extra five days, they finally just declared them feast days and tacked them on to the end of the calendar. But this did not really solve their problem entirely. It still left their civil calendar out of sync with the solar calendar and it would take 1,460 year cycles to bring civil and solar calendars into agreement.[1] This centuries-long cycle is called the Sothic cycle and some scholars estimate that the cycle may have been recognized as far back as 2850 B.C. But once the Egyptians finally devised an accurate calendar, they apparently lost all interest in advancing the science of astronomy. [2] They were much more concerned with verifying established data and observing the heliacal risings of Sirius. By this time, they had already lost interest in lunar phases and movement, for they had sundials to keep the time of day and water clocks to track time at night.

The sundial is an instrument that measures the time of day by a shadow appearing on a marked surface; the marked surface is divided into parts that represent hours or parts of an hour. In the very earliest periods, pyramids were probably used as sundial gnomons; that is, as the shadow-casting part of the sundial. This probability of pyramids also functioning as a gnomons is one of the reasons scholars assume the Egyptians had calendars as far back as the Old Kingdom.

Fig. 1-1. The sky to the ancients was both teacher and prophet. They were fascinated with the dependable motion of the planets and the dependability of the stars. This photo represents a wide-angle view toward the center of the Milky Way, taken by Dr. David Talent at the Cerro Tololo Inter-American Observatory, Chile (Courtesy of National Optical Astronomy Observatories).

**Table 1-2. Egyptian Cosmology.**

| Name | Description | Comment |
|---|---|---|
| Nun | Primal ocean representing a universe of chaos | This infinite ocean contained the basic constituents of all that would ever be. To Egyptians, water was the basic stuff of life. |
| Ra | The Sun God | Existed in Nun but remained in repose until he willed himself to life. From him came the air which holds up the sky, and the dew and the rain which moistens the Earth. From his tears, men and women were created. |
| Shu | The God of Air | Born of Ra, holds up the sky. |
| Tefnut | Goddess of Dew and Rain | Born of Ra; gave birth to Geb (Earth) and to Nut (Sky). But Geb and Nut married without Ra's approval and so he ordered Shu to separate forever Earth and the sky. |
| Osirus | God of Nature and Vegetation | First son of Geb and Nut and to him the Earth owes its fertility. |

The water clock (clepshydra) was constructed from a container designed in such a way that water would pass through it and into a second container or area at a steady rate over a 24-hour period. The second container was marked with at least 24 lines from bottom to top, each line representing an hour of the day. There could have been 48 lines in some cases so that half-hours could be determined—and quite possibly there were water clocks that were marked to measure quarter-hours. Sometimes these water clocks were simple vessels shaped like flower pots that emptied water at a consistent rate; and as the water level in the pot was reduced, lines representing parts of the day were revealed. Of course, someone had to keep watch over the water clocks so that they could be refilled at the right moment and no interval would occur that would result in incorrect measurement. Priests were usually assigned to stand the watch.

There was a third means of keeping time in those ancient days: the dehanal system. This method of keeping track of time necessitated dividing the celestial equator into 36 equal zones or dekans. It was expected that during the course of the lunar year, the 36 zones would rise in the east shortly after sunset every ten days. The method was used as an alternate method for keeping track of time at night. But again there was the problem of extra days with no place to go; 36 just does not go evenly into 365. So, the priests in charge of this time-measurement system had to do a little fudging when they made up their time charts. Using their special math, the dehanal charts they prepared actually accounted for the extra five days created by the 36 zones. Time and date could be determined by anyone who could use their tables in conjunction with the dehanal observed above the horizon.

Sirius (Fig. 1-2), known as the "Dog Star" or the "Nile Star," as well as by other names referring to its luminosity, was important to the Egyptians because it was used to forecast the rising of the Nile's waters, by far the most important event in the life of an Egyptian living near the river. Though it is 550,000 times more distant from the Earth than the Sun, Sirius remains an extremely brilliant star to observers, shining white with a tinge of blue—though it often gives the appearance of twinkling in all the colors of the rainbow.[3] Actually, in ancient times, Sirius might have been a different color than it is today, for, in the third century B.C., Aratus, the Greek statesman and general of Sicyon, described the star to be reddish in color, as did the Roman philosopher, Cicero (106-43), and the poet, Horace (65-8). So did other writers long before them, including Homer, one of the principal figures in ancient Greek literature and who probably lived in the eighth century B.C.[4] But if there has been a change in Sirius' chemistry so that in a mere 3000 years it has gone from a reddish-brown color to a bluish white, the star is on a remarkably fast evolutionary cycle,

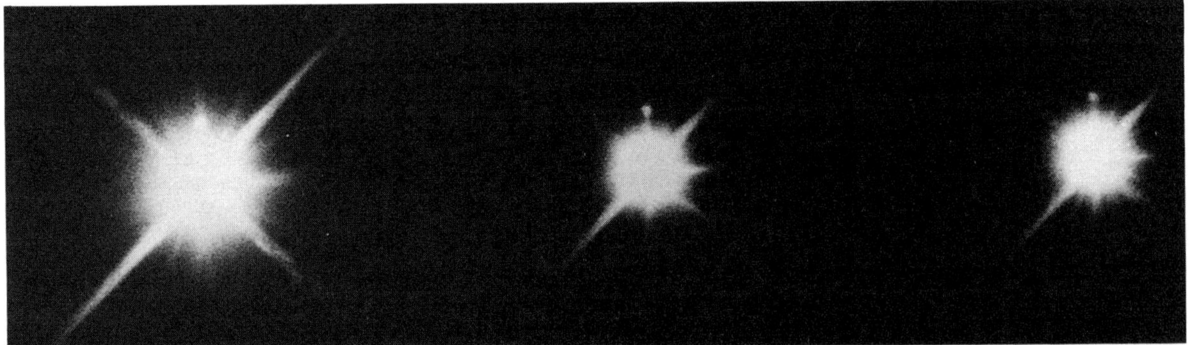

Fig. 1-2. Here are three exposures of Sirius, the Dog Star, taken with the 120-inch reflector at Lick (note the companion). Sirius is the brightest star in the constellation Canis Major and was probably the most important star in the sky to the ancient Egyptians, as it heralded the rising of the Nile (Courtesy of Lick Observatory).

one that might require a re-examination of what astronomers believe about the life-cycle of stars. Such a change in color represents a temperature change of thousands of degrees as well as a new mix of primary elements. Such temperature and chemical changes should take tens of thousands of years, not simply a few thousand years.

It has been suggested by some researchers that perhaps the ancients saw, interpreted, or described colors far differently than we do now, or that Sirius might actually have been in what is called the *red giant* stage a mere 2000 years ago when Cicero lived. The red giant stage is an evolutionary cycle that our own Sun will one day enter before it finally explodes into oblivion—but the Sun has about 5 billion years to go before it becomes a red giant. If Sirius was indeed in this red giant stage in Cicero's time, its immediate future, by stellar time, is relatively bleak. It will finish its life within the next 100 thousand years, passing into oblivion with a supernova explosion that hurls all of its material to unknown parts of the universe. Where Sirius now shines there will be only a massive cloud of gas. (Star evolution, chemical makeup, and color are discussed in Part 2 of this book.)

While Egyptian astronomy made little advancement after the second millennium B.C., the Mesopotamian civilizations continued to contribute heavily to astronomical knowledge and are credited with having founded astronomy as a science (Fig. 1-3). Mesopotamia was not an empire or a country but rather a geographical area in which people from various stocks settled and eventually organized city states, many of which date prior to the earliest Egyptian political communities. Most of the people of many of these Mesopotamian city states were Sumerian. The Sumerians had an extreme fascination with astronomy and astrology, and they pursued astronomy not simply for its practical uses as did the Egyptians, but also to acquire knowledge for its own sake.

The Sumerians came from, among others, the lands of Akbad, which had its beginnings with Semitic tribes dating back to the fourth millennium B.C.; Eridu, near the Euphrates in South Mesopotamia, which dates even further back (to the sixth century B.C.); Ur, the home of Abraham, and a land which flourished until about the time of Homer; and Lagash, a city which archaeological digs have revealed to be one of the most creative environments of those ancient times, which flourished until about the same time as the fall of Egypt's Old Kingdom around 2500 B.C. The names of these cities are lost to all but biblical scholars and a few serious students and teachers of history.

The Sumerians devised a calendar much like the Egyptians', based on a 360-day year. (The Babylonians used a similar system; they probably copied the Sumerians.) The Sumerian calendar, however, was different because it had provided for periodic adjustments to compensate for the fact that the year was actually 365 ¼ days rather than 365, and the lunar month was actually 29 ½ days rather than 30. This meant that their annual calendar would miss the mark by as much as eleven days each year. Thus, every few years, depending on the required adjustment, the Sumerians would add an additional month to their calendar.

The empires of Babylonia, Chaldea, and Assyria also extended over the Mesopotamian region.

Babylonian history is so inter-mingled with Sumerian and Chaldean history that its past easily becomes confusing. Some scholars trace Babylonian history to the fourth millennium B.C.; others trace it back only as far as the eighteenth century B.C. when Hammurabai established the first Babylonian dynasty. Much of the Babylonian educational system, writing, and science, however, had strong ties to the Sumerian culture; and Babylonian astronomy and astrology had its roots in Sumerian science.

Babylonian astronomy was left in the hands of an educated few who served as scribes and could use and understand the writing system handed down to the Babylonians by the Sumerians. This system of writing, which used wedge-shaped symbols instead of alphanumeric characters, is called *cuneiform*, and is the earliest known system of writing (Fig. 1-4). From early ancient to late ancient times, the symbols went through a graphic evolution and definitive evolution. As centuries passed, the sym-

Fig. 1-3. The ancients had to study the sky without the aid of lenses. There were no telescopes in their time. At best, they could see 6,000 to 8,000 stars, all of course existing within the boundaries of our Galaxy. Here is a photo of the sky with the Kitt Peak National Observatory 4-meter telescope. It takes us way beyond our galaxy to a cluster of galaxies located in the constellation Virgo. Thousands of galaxies make up these bright clusters (Courtesy of National Optical Astronomy Observatories).

bols not only took on new meaning but had drastically altered design. In their very earliest forms, the cuneiform symbols identified mainly physical objects but later on, symbols were added to represent abstract ideas.

Cuneiform symbols were drawn on clay and the clay was baked by the Sun's rays, or else in kilns, to make the writing as durable as possible. Fortunately for historians, the Sumerians, then Babylonians, were successful in finding a durable way to record their graphics; otherwise the thousands of cuneiform documents available in museums and libraries throughout the world would not be available for study today. Cuneiform deciphering, however, is a relatively new art, for it was only in the beginning of the nineteenth century that a German high school teacher was able to find the underlying logic in the wedge-shaped symbols. It was not until the year 1846 that the first translations were made available, made and published by an Englishman, Sir Henry Rawlinson. Actually, Rawlinson first deciphered Persian cuneiform; he quickly moved on to Babylonian and thereby opened the curtain to a new stage in history. Astronomical data of Babylo-

**Table 1-3. Babylonian Cosmology.**

| Name | Description | Comment |
|---|---|---|
| Apsu & Tiamat | Goddess of Sweet Water, Goddess of Salt Water | Water was the primary element in the universe. Everything began when Apsu and Tiamat united. |
| Ea | God of the Waters | Great grand offspring of Apsu and Tiamat; rose against them when he learned of their plans to destroy generations of gods. |
| Marduk | God of Fertility | Called upon by Ea to help destroy Tiamat, who was much too wily and fearsome for Ea to handle. When Marduk killed Tiamat he made the dome of the heavens with one part of her body and the Earth with the other part. |
| Kingu | General in Tiamat's army | From his blood, Marduk created men and women. |

nian origin was first deciphered in the 1880s by two Jesuit priests, Johann Strassmaier and Joseph Epping. These two dedicated scholars revealed that the Babylonians had mathematically described lunar motions, knew and named the known planets of the time, and had given names to the constellations which made up the zodiac. Detailed studies since have revealed that the Babylonians had little interest in pursuing observational astronomy to greater heights but seemed, rather, to be preoccupied with putting to mathematical formula known phenomena. What advantage would there be, they must have asked, in understanding orbital patterns or the effect of one celestial body on another when the stars and planets were so far beyond reach that such knowledge would represent an exercise in futility? It was much more to the Babylonian advantage to take a numerical accounting of past astronomical events, formulize them, and thereby be able to predict, or at least expect, the recurrence of special celestial events that had an impact on their daily lives or which they could witness without fear—for instance, lunar eclipses.

The main concern of the Babylonians was always the precision of their calendar, so they were necessarily preoccupied with lunar and solar phases. It had always been a Babylonian custom to designate the first day of each of the twelve months as the next day after the appearance of a new moon. For centuries they determined their monthly cycles by constant observation but in 400 B.C. they had effectively put into formula the relatively changing speeds of both the Sun and the Moon when they were moving from west to east around the zodiac. At first the Babylonian scribes established two different speeds for the Sun and Moon so that the average of these speeds could be used to calculate the true zodiacal movement. As time went by, however, their math became more sophisticated until finally they developed the linear equations that effectively represented the minimum to maximum speeds in one direction and the maximum to minimum speeds in the other direction.

Babylonian mathematics would be a bit cumbersome for those of us who are used to the decimal system. We count everything in powers of ten, using ten digits from zero to nine to represent units, and using positional notations to represent the powers of ten. The Babylonians used a base-60 counting system which, by the way, led them to divide the circle into 360 degrees and the parts of an hour down to seconds in a sexagesimal measure. Thus, there are 60 seconds to a minute, and 60 minutes to an hour. Fortunately, when they decided

Fig. 1-4. Cuneiform graphics represented a system of writing developed about the fourth millenium B.C.. The Sumerians are generally credited with having originated the system. Cuneiform means "wedge-shaped" (Courtesy of Library of Congress).

to divide the calendar into smaller segments—namely, the week—they did not try to find some fraction of 360 to represent the number of days and were willing to settle on seven.

Many of the Babylonian ideas in astronomy as well as astrology were influenced by the Chaldeans. The Chaldeans were from the southernmost section of the Tigris and Euphrates valleys, although sometimes their political boundaries extended to include all of Babylonia. They invaded the region during the eleventh century and are credited with restoring what had become a dwindling Babylonian empire. The Chaldeans are remembered in the history of astronomy more as astrologers than astronomers.

Even before they came to dominate Babylonian society—for they were at first but a small kingdom within the Babylonian empire—the Chaldeans were recognized as zealous worshipers of the sky and also as highly skillful magicians. The word magician, it should be noted, had a different meaning in ancient times than it does now. Today, we think of a magician as a trickster, a person who can fool us with all kinds of illusions. But ''magician'' in the days of the Chaldeans, the Babylonians, and the Magi referred to someone who knew enough about the natural world to heal the sick, understand and use natural forces, explain the movements in the sky, and explain and use the science of the times. Magicians in those days were prophets, soothsayers, psychologists, and compilers of knowledge. The Chaldeans were a learned people. They knew the Earth, the stars, and the history of the Mesopotamian and neighboring peoples. They could recreate the past with their vivid histories, foretell the future from their knowledge of astrology. Their name became synonymous with magician and astrologer.

The Chaldeans were always involved with trade of one kind or another throughout and beyond Mesopotamia, with the Egyptians, Assyrians, Persians, and Greeks. They were quick to adapt that which was of practical use and were quick to learn from all who would offer them new ideas, methods, or technology, but in their religious traditions and in their science, they remained primarily Babylonian.

Chaldean cultural dominance continued throughout Mesopotamia until the Assyrian conquests, which began with greater force than before, during the eighth century B.C. The Assyrian assault on the Chaldean empire continued into the seventh century, by which time Chaldean and Babylonian cultures were so intermingled they could not be distinguished.

The Assyrian empire had its beginnings sometime around 3000 B.C. The people were Semitic and for the first 1800 years they were dominated by the Hittites and/or the people of Babylon. In the twelfth century B.C., however, their numbers, military might, and political ambitions swelled to such a level that they chanced a series of campaigns designed to bring all of Babylonia under their domain. Their dreams went unrealized for centuries, for three hundred years would pass before they began to dominate the Babylonians. By the reign of Ashurbanipal (668-626), the superior iron weapons of the Assyrians led them to defeat the Palestinians, Phoenicians, Armenians, and Medes.

Fortunately, the Assyrians had a tendency to preserved and remained available as an accurate quered territories. Thus, the Assyrian empire became a melting pot of art, science, and culture. They copied and studied Babylonian literature and science, became fascinated with Chaldean astrology, and drew heavily from the political and religious ideas of the Medes. Once introduced to Chaldean astrology, however, the Assyrians could not let it go; it fascinated them and entrapped them. Now, when the Assyrian astronomers looked to the heavens, they cared little about acquiring new observational evidence or the formulas which drove the heavenly bodies; they cared now only for the religious, mystical, and prophetic nature of the sky. They believed even more than the Chaldeans that the future was written in the stars.

Chaldean astronomical tables were, therefore, preserved and remained available as an accurate source of reference to future researches, such as Hipparchus and Ptolemy. The Chaldean tables were ingenious. For instance, consider the record of observations and the resulting calculations for establishing something called the saros cycle (the period in which eclipses of the Sun and Moon repeat themselves). By the Assyrian conquests, the Chaldeans had already established the interval to be 6,585 ⅓ days, or the equivalent of 223 lunations (synodic periods).

The Assyrians were highly superstitious, whether or not they were religious. The average Assyrian soldier who cared little for the priests and their religions still looked for approval from the gods and believed in the physical embodiment of evil. He also looked for omens for everything he did. And just as each Assyrian soldier feared and respected the supernatural so, too, did the Assyrian rulers. They wanted as much to appease the gods as their most humble servants. They sought supernatural signals for each and every personal, political, or military adventure. It was important for them to know if there would be divine intervention in any of their undertakings and whether this intervention would be for or against them, for certainly the gods would take sides in any dispute or interfere with any undertaking if the results would not be to their satisfaction.

Men and women who could read the omens from Earth and sky, therefore, were in special demand. Kings, queens, princes, and those in political office or military command paid handsomely for their advice. This was one of the reasons the Chaldean priests were able to wield so much influence. Their ability to interpret the hidden meanings behind all natural events, including the signals apparent in the motion of the planets, made them feared and respected. But the task for anyone who dared claim to know the omens was a dangerous one. These were bloodthirsty times. Pure guesswork that did not pay off could result in the magician's execution. Thus, a great deal of creativity was used in forecasting events; astrologers realized the advantage in using a syntax that could be interpreted in a number of ways and therefore left the action to be taken up to the individual petitioning the astrologer. Only someone knowing his own soul could decide the correct way in which to interpret the omens. However, there were times when astrologers seemed willing to throw all caution to the wind and go all-out in their predictions. Perhaps, they were the novices or the ones who were riding on a great history of success—or, perhaps, they were no longer in a position to double-talk.

But while the Chaldeans and the Assyrians became noted for their magic and wisdom, there was another tribe of people who had studied the stars with the same enthusiasm and religious purpose as the Chaldeans. They had become famous for reading the will of the divine and for being able to foretell the future. They were the Magi, one of the seven tribes of the Medes.

Now, the origin of the Magi is a mystery. There is mention of them in biblical manuscripts and also in other ancient documents.They were not of the same stock as the other tribes of the small kingdom of Media. They were generally considered with some prejudice. Their magical powers and knowledge of the heavens, however, had helped them gain a political foothold, and as time passed they became the keepers of the Median religion. Their belief in human sacrifice also helped their political cause. If someone got on the wrong side of them, he or his relatives might be the next sacrifice.

Media was located in what is now western Iran. The Magi might have emigrated from the less-civilized eastern sector of Persia (which was the name for Iran from ancient times until the 1930s). Or they might have been members of the Chaldean or Babylonian empires who found it to their advantage to emigrate and form their own community, which was eventually joined to the other tribes of Media. If the origin of the Magi remains an uncertainty, their reputation as soothsayers, prophets, magicians, and astrologers had no uncertainty about it. Their image as great intellectuals, skilled in the arts and sciences as well as in magic, probably reached its height around the fifth century B.C. and continued near its summit until about the third century B.C.

One of Plato's greatest wishes was to one day be able to travel to the land of the Magi and study their arts and sciences. Destiny would deprive Plato of this privilege, for Grecian political and military conflicts, as well as Plato's own involvement in teaching and writing, prevented him. One of Plato's students, Aristotle, did visit with the Magi and eventually compiled a book, now lost, on their society, their culture, and their knowledge.[5]

The Magian interest in astronomy, as great as it may have been, eventually went the way of the Chaldean interest; that is, the Magian priests became so preoccupied with the art of astrology that they laid aside the science of astronomy. Astrology became as much a part of their religion as the teachings of Zoroaster eventually did.

No other people in these ancient times were as famous, or more famous, for their astrological abilities than the Magi.[6] They had little patience with those who aspired to other religions. Although they worshipped fire, they were not idolaters. Fire to them was simply a symbolic figure of a god so existential that man could never reach him, or be equal to him. Magian religious principles were adapted by the other Median tribes—or, perhaps, forced on the other Median tribes. In its philosophical foundation, the Magian religion was not unlike the Jewish beliefs of the time. To both the Medes and the Jews, there was only one God and he sought the highest good from his people. And both the Jews and the Medes expected a redeemer, a messiah who would lead their people to glory; only in the case of the Medes, particularly the Magi, the idea of a redeemer only came after they had adapted Zoroastrianism, which expected three messiahs instead of one, the second and third coming after one thousand years. Magi and Jew, then, had the same fundamental expectation. But the Magi did not know when the first messiah would come, so they hoped for some special sign. Being obsessed with astrology, they assumed that the sign would come from the heavens. Some planet would change its course, some celestial display would occur, or perhaps some star would shine brighter than all the others and lead them to their redeemer. So the Magi lost all interest in developing their astronomy to new heights; they only cared now about how better to read the messages in the sky.

While the Chaldean and the Magi are the ancient peoples most closely associated with the subject of astrology, the subject itself might be traced back to back to the eighteenth century B.C. and to Babylon when the cuneiform text *Enuma Anu Enlet* was fixed in clay. This text concerned itself with celestial omens and how they might be interpreted, and it was probably the forerunner of the more standardized astrological schemes developed after the sixth or fifth century B.C. In the *Enuma*, there were four major deities with whom the listed omens were associated. Of these four, three were related to celestial objects: the Moon, the Sun, and Venus. The *Enuma* had sufficient scientific foundation to hold up its theoretical proclamations and therefore it was widely received and respected by other cultures right through the ancient period and into the medieval. Ptolemy, who as a matter of style liked to record or verify the works of those before him, copied the Enuma in his *Tetrabiblios*, which he had written in the second century A.D.

Other texts besides the *Enuma* concerned themselves with celestial and other omens. These were written with such good argument based on scientific fact (as, for instance, the very predictable movement of the planets) that astrology came to be looked upon as truly "the science of the stars." But what astrology did in actuality was to inhibit the further development of astronomy as a science. It severely handicapped observational and theoretical progress and left astronomy a tool for the religious apologists and political manipulators. Once the ancient religions and astrology became so interdependent, there was little room for the startling new astronomical or cosmological theories that might have laid a very different foundation for medieval astronomy. Once the science of the stars came to have serious religious implications, even the more outspoken scientists/philosophers of the time rarely dared to challenge prevailing beliefs. In those days, religion was not something one carelessly spoke out against. Besides, the more enlightened generally felt that it was not worth challenging the accumulations of ideas and practices of established religions or social customs that were so imbedded in the average citizen's way of life.

The cosmogony (study of the origin of the universe) and cosmology (study of the universe as an orderly system) that did develop in these ancient cultures had roots not in astronomy but in religious mythology. Cosmogony actually came first, then cosmology, then astronomy and astrology. In all cases, both cosmogonical and cosmological ideas presented by the Egyptians, the Babylonians, the Assyrians, the Persians and even the Greeks were relatively unsophisticated, particularly when compared with the achievements of these same cultures in the area of astronomical knowledge. Although the cosmological and cosmogonical schemes of these ancients were simple in their construction, they had a lot more to them than is immediately apparent.

The Babylonians believed that everything originated with water. Before the earth, the planets, and the stars, there was only a vast cosmic ocean which contained the seeds for all that would eventually be in the physical universe. There was another eternal ocean from which our universe was born; this second ocean was the one on which the heavens floated. No one could make the journey from the one great ocean to the other—except, perhaps, by the will of the gods.

In Egyptian mythology, also, before the creation there was only an eternal ocean. This great ocean was called by the Egyptians the "great void." There were inorganic, eternal dwellers of the great ocean which were like tiny seeds which contain the potential for becoming greater things; but these "seeds" in the great void do not have the breath of life to give them direction and purpose until some moment in eternity when one of those seeds willed itself into existence and became the sun-god, Ra. Ra, in turn, created men and women and all the world. From this cosmological scheme, the Egyptians created a universal model that had Earth as the center of this great aquarian universe, an idea also adapted by Assyrian cultures.

Persian cosmogony and cosmology is probably more closely associated with Zoroastrian ideas brought to the Persians by the Medes, particularly the Magi. There is, however, some connection between the Zoroastrian and Egyptian cosmogonies. In the mythology of both peoples, the first god to come into existence did so simply by willing it. Zoroastrian and Egyptian deities, therefore, actually preceded themselves. Somehow, to them, *will* was an eternal force that existed regardless of life. But while the Egyptians' sun-god emerged from the great void, the Zoroastrians (and, therefore, the Magi, Medes, and Persians) believed that Oramzd willed himself into being from absolutely nothing at all. Because the ancients knew that everything they saw or experienced came from something, they could not conceive of something just happening. Somewhere at the beginning of the causal sequence there must have been a will involved somehow, someway. There must have been something that always was, something that preceded itself, something that actually willed itself. It was an astonishing idea to them but no more astonishing to them than the idea of something coming from nothing. Given the alternative of accepting the idea of something coming from nothing or something always having been in existence, they opted for the latter. In each case—that is, for the Egyptians, the Babylonians, the Assyrians, the Medes, the Persians—this ever-existent entity was a god *who willed himself into existence*, meaning he existed somehow somewhere prior to that moment. Their gods were pre-existences which came to be. They could think of no better explanation for the reason for the universe. Why did they not just assume that matter always existed and the world was born from some strange catastrophe—or, perhaps, had always been, perhaps as a seed at one time in the great void? It crossed their minds. But there was intelligence in the universe, and they could not conceive of intelligence born of some inorganic substance with no will of its own.

The idea of a great primeval sea in which all the seeds of existence floated aimlessly is not quite as far out as it sounds. Today, physicists see space not as an empty area between points or objects but rather as a medium which connects these points or objects. In Einstein's spacetime physics, space expands, contracts, curves. It is the road we travel to get from here to there. As a medium, space can be compared with a great ocean, the waters of which can be navigated to bring a traveler from one location to another, or some great object to another. To the ancients, the planets and the Moon seemed to float in the sky much like objects contained on a great sea.

In current cosmological models, the universe is often seen as a great balloon speckled with dots of galaxies. The interval between these dots represents space. As a balloon can be expanded or contracted, so too can the universe expand or contract during its evolutionary phases. Current theories have it that once there existed nothing at all or, perhaps, just some primeval atom. There was only a great void and perhaps this one single existence. Then suddenly there was a great explosion of sorts and the primeval atom—or the suddenly-created atom—began to expand and expand like a balloon. Eventually stars and galaxies formed within its parameters and even the space between and within these galaxies. This universal balloon is still expanding, according to current theories, with the galaxies receding further and further from each other while the space between the galaxies stretches and stretches and stretches (Fig. 1-5). So, you see, the ancient cosmologies were no less strange than those we theorize about today. Once there was a great void and then a big bang, some say today. Once there was a great void and then a god who willed himself into existence, the ancient Egyptians say.

The Zoroastrians believed that Oramzd created a great fire from which all the world was created. (The ancients used the word "world" as we use the word "universe.") One of the current cosmological models explains that the universe was created by a great fireball which set in motion the series of causes and events which brought about the stars, galaxies and planets. Scientists today offer, among others, these four theories about the origin of the fireball:

- It just happened, nothing preceded it, and all matter was created at the same time as the explosion which generated the fireball[7]
- The explosion occurred when an area from a previously existent universe was involved in some catastrophe
- Before all of creation as we know it, there existed some primary particle which, for unexplained reasons, exploded with such force that all its pieces were hurled out in all directions at once, expanding and receding as they raced away from the nucleus
- A supernatural being started the fireball, and therefore all of creation, by so willing it.

The Persians and the Medes would have gone with the first explanation, for it is little different from their own. The Egyptians, Babylonians and Assyrians would have liked the third, for it would fit in with their own ideas.

## THE LATE ANCIENT PERIOD

Astronomy and astrology developed independently across the Eurasian continent in China. Yet there are some similarities between the European and Chinese science of the stars that leaves one convinced that there are serious gaps in the panorama of ancient events. Frequent interchanges between the West and the East during the 1200 or so years from the sixteenth century B.C. to the third century B.C. seem to be indicated because there are striking parallels between Chinese ideas and those of the Mesopotamian, Persian and Greek cultures. However, China was on numerous occasions raided by barbarian tribes from the north and the west; these tribes may have had direct contact with the peoples of the western areas, either through trade or conquest.

The Yellow and Yangtze river basins formed the heart of ancient Chinese civilizations. The earlier settlements expanded steadily beyond those basins, with occasional periods of contraction as foreign invaders swarmed victoriously upon Chinese cities. Most of early Chinese history is actually a story of invasion after invasion, cultural smothering, and then successful revolt and cultural expansion. Chinese astronomy probably had its start with the cultures that emerged in the third millennium B.C.; the sources referring to the beginnings of Chinese astronomy date from the Shang and Chou dynasties, which ran

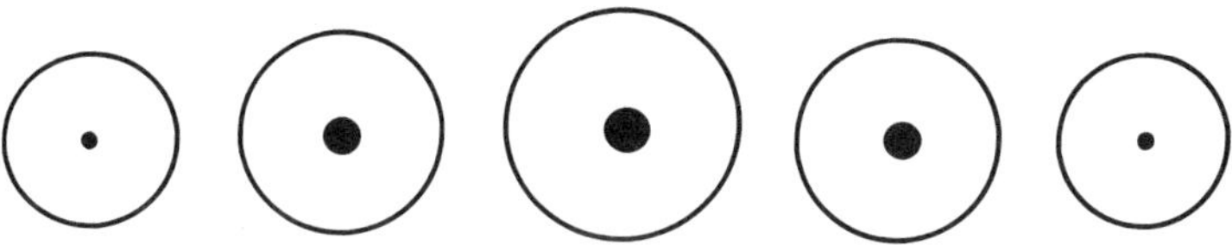

○ = Space expanding & contracting

● = Solar system expanding & contracting along with space

Fig. 1-5. We may think of space as an empty field between objects, but in physics space is an elastic medium that can expand and contract.

**Table 1-4. Late Ancient History of Astronomy (700 B.C.-200 A.D.)**

| Astronomer/Philosopher | Comments |
|---|---|
| Thales (624-547) | Possibly compiled a work on navigational astronomy; developed astronomy as a science; believed the principle of all things was water. |
| Anaximander (610-546) | Developed a theory of natural evolution; believed there were fundamental forces governing all of nature; believed space to be finite; Earth is 1/20 the size of the Moon; Earth is cylindrical with a depth 1/3 its breadth. |
| Xenophanes (579-478) | The Sun travels at an almost infinite distance from Earth, thus the reason it appears to set; the Moon is the source of its own light; eclipses are the result of the Sun's light burning out ahead of schedule. |
| Pythagoras (570-500) | Believed all the world (universe) could be defined in terms of the four critical numbers; the universe is constructed on ten spheres; the celestial order is Sun, Mercury, Venus, Mars, Jupiter, Saturn; Earth is not the center of the universe and is only an insignificant object in the universal scheme of things. |
| Anaxagoras (500-420) | The Sun is a burning world of stone; the Moon reflects light from the Sun; planets began as small particles that crashed together to form larger bodies; there are infinite principles governing the things in nature. |
| Leucippus (Fifth Century B.C.) | There are an infinite number of elements existing in eternal space; scientific theories carried on by his friend and student, Democritus. |
| Democritus (460-370) | All existing things are made up of atoms which are in perpetual motion; Earth, planets, and stars were formed as a result of the acceleration of minute particles. |
| Plato (427-347) | The heavens are divine perfection; by studying planets and stars men can learn more about agriculture and navigation and learn the proper method of scientific investigation; all movement in the universe is caused by soul; Venus is closer to the Sun than Mercury; the heavens are only a mirror of reality. |
| Aristotle (384-32) | Starlight is produced by the heat and motion of the stars; stars are spherical but neither spin nor roll; the Sun, Moon, and planets consist of a necessary substance called the ether. |
| Eudoxus (408-355) | Built his own astronomical observatory; determined the length of the solar year; developed his own calendar; explained the position and movement of the stars according to scientific principles. |
| Heraclides (388-315) | Earth rotates and the planets orbit the Sun; the stars and planets move in concentric and eccentric revolving spheres. |
| Berosus (3rd Century B.C.) | Developed a sundial called the hemicycle; the Moon is a sphere only half of which gives off its own light; Earth is ½ million years old. |
| Aristarchus (320-250) | The Moon shines by reflected light; the Sun is 18 times as far from the Earth as the Moon; the Sun is the center of the Earth's orbit. |
| Erathosthenes (275-195) | Determined the approximate size of the Earth and the approximate distances from the Earth to the Moon and to the Sun. |
| Hipparchus (190-120) | Recorded the positions of the stars and the movements of the planets; was the Tycho Brahe of the late ancient period; argued for the heliocentric theory; produced a catalog of 1000 stars. |
| Posidonius (135-51) | Compiled a great deal of data on Earth's atmosphere; produced remarkably accurate measurements of Earth; made some of the best estimates of the Sun's diameter and its distance from Earth. |
| Lucretius (98-55) | Not a scientist but a poet/philosopher whose ideas about comets were far ahead of his time; the universe is an infinite void consisting of invisible particles. |
| Ptolemy (100-170) | Synthesized, verified, and expanded ideas of earlier astronomers; drew heavily from Hipparchus; believed Earth to be the center of the universe. |

concurrently from about the sixteenth century B.C. until the third century B.C. The Chou dynasty represented a series of successive reigns that were marked by many wars and internal political conflict. Toward the end of the Chou dynasty (1027 to 256), China entered its first great period of artistic and philosophical glory. The period following the Chou dynasty was the time of Lao-Tzu (604-574?), Confucius (551-470), Chuang-Tze (fourth century B.C.), and Mencius (372-288).

Lao Tzu (Lao-Tze) was the founder of Taoism, which was both a philosophy and a religion. Actually, there is some doubt about whether or not Lao-Tzu is indeed the founder of Taoism, for many scholars have judged that the text call *Tao-Teh-King*, which sets down the Tao tradition, was written some 300 years after Lao-Tzu. The Taoist philosophy elevated contemplation and taught that all men should seek to free themselves from the temporal order and the sensual things which relate to it—at least this is the way the doctrine was interpreted by Chuang-Tze, the mystic who idolized Lao-Tzu while at the same time criticizing many of the ideas of Confucius.

Confucius (Kung Fu-Tse) is himself a first-class oriental mystery, for, despite his impact on Chinese thought, little is known about him personally. There is a bit of a parallel that might be drawn between Confucius and Socrates in that as we only know about Socrates from dialogues written by Plato, so we really only know about Confucius by a series of dialogues compiled by his students. And, as Socrates did, Confucius traveled about with students and disciples constantly at his heels. He may have been the author of, or in some way was intricately involved in bringing about, the *Wu Ching* (Five Books). This text consists of the *Shu Shing*, the *Ch'un Ch'us*, the *I Ching*, and the *Li Chi*.

The *Shu Shing* reaches back into Chinese history to tell of the old and important customs that should be honored; the *Ch'un Ch'us* is a history of Confucius' homeland (Lu); the *I Ching* is a mystical piece dealing with supernatural quests and bequests; and the *Li Chi* is Confucius' "Utopia." The religious philosophy taught by Confucianism developed not from the Five Books but from the *Lu Yu*, analects of Confucius contained in the *Shih Shuh*. The Confucian philosophy stressed the social consciousness of mankind and taught that the interaction of all men and women could be shown by the primary relations: between the rulers and the ruled, between parents and their offspring, between elders and youngers, between husbands and wives, between friends. Confucianism sought harmony in all relationships, not only between men and women but between all of nature and each and every man and each and every woman. Here were the roots for a communion between man and the stars and the increased regard for omens of any kind, for if men and women were in communion with the rest of the natural world, the world could speak to them, as they to it. (Confucianism was a strange combination of moral orders, for it combined philosophy and religion without including theology as an intermediate order between the two. One would expect that religion, for instance, would require a deity which sought communion with its congregation. Confucianism was never conceived to be a theology, though by the second century B.C., its followers allowed that indeed there must be a higher conscious authority than the reigning sovereign so the Confucians came to worship a god named Shang-Ti.)

Mencius, the last of the great Chinese philosophers mentioned here, argued that man was a compassionate and social being rather than a self-serving animal. In its extension, his philosophy had severe political implications, for he taught that when the political institutions bring men and women to fulfill their selfish passions, then the sovereign responsible for the decay in national character should be removed in whatever way necessary. What Mencius had tried to do was develop a political and economic structure from fundamental Confucian philosophy. He believed that the responsibilities of a sovereign to his subjects overshadowed the responsibilities of the subjects to the sovereign. That is, it was of primary importance to the people at large to find a communion with their fellow men, to live with compassion for each other, and to develop a social order that was of intrinsic value; and it was the sovereign's primary responsibility to see that the political and social order which he maintained was conducive to such an ideal social and political existence.

Though Chinese philosophy had, in this late ancient period, recognized a higher order than that to be found on Earth, the Chinese had great difficulty in explaining the how and why of the deity. If there were a god or gods, the Chinese could never be certain about what it expected of them or what they might expect of it. Thus, they never became preoccupied with establishing direct communion with their deity or deities, and they seemed to believe that the deity left the physical world and the institutions of men to man himself. The gods did not determine the fate of men and women directly. Rather, the fate of each individual was determined by their adherence to the highest possible good, for in seeking the highest good, each man and woman sought what was ultimately best for themselves. To the Chinese, virtue was the highest good and those who knew most about virtue and what was best for mankind were the students and teachers of history, the wise men who knew the ancient documents and poetry and, therefore, were closer to the source of all knowledge. The ancient Chinese were obsessed with the past; the more ancient an idea or a document, the more beautiful, the more correct, the more valuable, the more wise. If the Chinese truly worshipped anything beyond this world, it was the memory of their ancestors, and the memory of the great philosophers and sages who taught in times past.

Thus, one of the advantages the Chinese sciences had over western sciences was that there were no religious traditions or prejudices that could be challenged by the objective search for knowledge. The Chinese were not raised on great stories of gods being born from aquatic universes, or deities that made man and his world the center of all the universe. The Chinese could develop any cosmological model they thought valid without fear of challenging the great mythologies like the Egyptians or Babylonians had developed. Actually, the documentation of scientific and astronomical inquiries and discoveries in China up until about the second century B.C. is relatively scant compared to the wealth of documentation produced by the Egyptians, et al. This is probably because of the cumbrous techniques the Chinese used for inscribing their ideas, using bamboo as a writing material. Once silk replaced bamboo in the second century B.C., the amount of documentation produced in the arts and the sciences increased steadily and dramatically. And by the beginning of the second century A.D., the last century before the beginning of the middle ages, the number of Chinese books and other documents increased even more dramatically—for now the Chinese had paper.

It is evident from the wealth of Chinese documents so far compiled that the need for astronomical observations came about because the sovereigns of the Chou dynasty wanted a very exact accounting of the passing of time. Efficient management depended upon being able to time things and intervals. Further inspiration for studying the heavens came about as a result of the age of intellectualism that began in the last two centuries of the Chou. These were times of great inspiration, and as new developments were taking place in the religious and political arenas, the Chinese citizen sought to understand all that was happening. He learned to read, to challenge what he read, and to look for solid logic in all that was taught. Confucius' writings introduced the idea of omens to the inquisitive citizen; in fact, one of those books was dedicated to little else than omens and mystical phenomena. It was only to be expected that as astronomers observed the heavens in order to perfect their calendars and other means of measuring time, they also sought to find the very special meanings hidden behind everything that happened in the sky.

Court or state astronomers played important roles in the Chou and later the Ch'in (227-201) and Han (202 B.C.-220 A.D.) dynasties. The Han dynasty was, perhaps, one of the most peaceful in China's history and the one which encouraged the highest intellectual achievement in ancient times. Even before the Han dynasty, astronomy had already become not only a respected science but a very necessary one. Astronomers became the sentinels of empires. They kept track of the days, months, and seasons; they prepared and published calendars; they mapped the constellations and read the stars so they could mark their calendars with days of fortune and days of misfortune. They told the emperor and his people the right times for celebrations, rituals, love, war, business, and any event of importance. The astronomer/astrologer had to be ready to defend his reasons for any prediction that eventually went sour, as did astrologers in the west. The emperor wanted only the best advice, advice that he could trust. The astrologer was his guardian just as were his personal bodyguards. An astrologer/astronomer suspected of being derelict in his duty, fraudulent, or incompetent would soon be sent to join his ancestors.

Chinese astronomers/astrologers realized the importance of their work and were proud of their responsibilities and titles. They were particularly careful to record all celestial phenomena. Even now, using Chinese documents, astronomers can track Halley's comet back to ancient times. By the Ch'in dynasty, the Chinese astronomers had mapped more than a hundred constellations and had catalogued more than a thousand stars. They had their own version of the zodiac for tracking the Moon and the planets, though they knew nothing of the motions of Mercury and Mars until sometime in the middle of the Han dynasty. Still, despite their increased knowledge of the heavens, the Chinese seemed little interested in developing a reputable cosmogony or cosmology. Most of the known cosmological schemes that have Chinese origin can only be traced back to the very end of the Han dynasty or to the early part of the middle ages. Like their western counterparts, the Chinese astronomers believed that the world was the center of the universe and all the heavenly bodies, including the Sun, orbited about it (Fig. 1-6).

Like Chinese astronomy and astrology, the Greek art and science of the sky began to develop around the seventh century B.C.; the Greeks were a rather late arrival on the ancient scene. This does not mean, however, that the geographical region on which the Greek city-states eventually flourished lay unsettled before the seventh century B.C.; the great Minoan and Mycenaean civilizations once prospered in these territories.

The Minoans inhabited an area in southeastern Greece near the southern boundaries of the Aegean Sea from about 3000 to 1000. The Mycenaeans were an Indo-European, Greek-speaking civilization that made their way into Greece shortly before the beginning of the second millennium B.C. To the Minoans lies the credit for the development of hieroglyphic writing; to the Mycenaeans lies the credit for salvaging the cultural contributions of the Minoan civilization when it began to fall into political, economic, and social disorder around the tenth century B.C.

The Mycenaeans eventually fell victim to the warring Dorians who, by the middle of the tenth century, had come to dominate Greece as well as to found a number of city-states which seemed more interested in fighting with each other than in contributing to the progress of Dorian culture. By the time of Homer the blind poet, sometime in the eighth or seventh century B.C., Greece had become a land of several politically independent people, among whom were the Achaeans, Aeolians, Ionians, as well as the Dorians.

Because they were great navigators and explorers, the Greeks were involved in continuous exchange with many of the ancient cultures on the African continent, in Mesopotamia and Arabian lands, and possibly also in the Far East. Besides

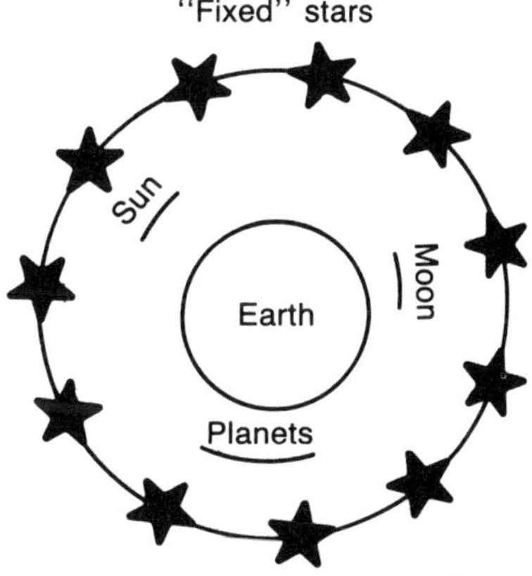


Fig. 1-6. It was hard for the ancients to comprehend any other cosmological model than one that had Earth as the center of the universe.

their continued interaction with countries to the west, south, north, and east, another notable characteristic of the Greeks was the freedom of artistic and scientific expression that they enjoyed. They had no cultural, political, or religious institutions which smothered innovation because it challenged tradition or the status quo. The small, almost always warring city-states had their own deities and their own priestly classes, but the gods were second and the priests third to what was materially best for each and every citizen.

Greek astronomy became as creative as it was technical. The Greeks let their cosmogony or cosmology develop in whatever direction intellectual pull took them. The Greek scientists and philosophers—in ancient Greece, philosophy and science were inseparable—could let their research and theories take them in whatever way they may.

The first Greek credited with introducing astronomy on strictly scientific grounds was a philosopher born in Miletus, then a seaport in Asia Minor. Remembered now as the Father of Philosophy and as one of the seven wise men of Greece, his name was Thales and he lived from about 624 to 547. (Some historians fix his birth at 640 B.C. based on background data found in the works of Diogenes Laertius, a third century A.D. Greek philosopher who produced an invaluable 10-volume work on the lives of the great philosophers who lived before him.)

There are many accounts of Thales' travels, and it appears that he spent some time in Egypt studying science and math. One result of his intellectual pursuits in Egypt was his carrying back to Greece the fundamentals of geometry. Thales lived in a time of great artistic and intellectual progress; but his time was also a time of numerous tyrannies, resulting mainly from the political inequalities between the landless and the landlords, the discontent of a peasant class reduced to slave labor or banishment, and the aggressive militarism of the slave population that was beginning to interfere with Grecian free enterprise. The result of all this discontent was a series of movements by the politically ambitious to over-

**Table 1-5. Greek Cosmology.**

| Name | Description | Comment |
|---|---|---|
| Chaos | Empty Void (or, meaningless existence) | Void had a very different meaning to the ancients than it does to us. It was not nothingness but disorder or meaninglessness. From Chaos was born Gaea and others. Gaea was Earth and the son to which she gave birth was Uranus (the Heavens). |
| Eurynome | Goddess of the Universe | Separated the sea and the sky and gave birth to Ophion. |
| Ophion | The Serpent | Impregnated Eurynome and thereby fathered the cosmos and all that exists in it. |
| The Titans | Gaea (Earth), Uranus (the heavens), Oceanus (the great ocean that surrounded the Earth), Hyperion (god of light, father of the Sun), Themis (god of Justice and Order); Prometheus (creator of men and women); and others. | They were given the planets to rule when Eurynome banished Ophion, who had tried to take her authority from her. |

(Greek mythology is complex. A number of different myths prevailed in ancient times. This summary represents a composite.)

throw established governments and set themselves up as tyrants in the many Greek cities—but never in Sparta, which had a political, military and economic structure too strong to be undermined.

These political upheavals worked to the advantage of the creative arts. Reigning tyrants, in an attempt to keep their subjects content, encouraged literary, scientific, and artistic efforts and institutions. Greek philosophy actually has its roots in these times, founded, of course, by Thales. But his school, the Milesian school, though remembered in histories as a school of philosophy, was also a school of science. Thales himself was an engineer and had once diverted a river to allow a royal parade to continue on its journey; and he was an astronomer, sometimes credited with predicting an eclipse which occurred in 585 B.C. Thales taught that man's soul had the power of self-motion and the power to start movement in other things. The chapter on light discusses how this philosophical foundation affected the way in which the early Greeks perceived vision; that is, as something which came from within, which was a power of the individual; a man or a woman's eye snatched the image of objects by emitting rays.

There are stories that Thales' preoccupation with the search for knowledge led him to forget his own economic welfare and so his detractors often found reason to publicly ridicule him, especially for his inability to use his intellectual prowess to build a fortune. However, he once struck back at those who laughed at him by proving that if he were indeed interested in simply material wealth, he could achieve it. He proved this by using his knowledge of astronomy to do a little commodity trading. Knowing one year that the stars predicted a fruitful harvest of olives, he rented all the available olive presses in Chios and Miletus and cornered a good segment of the market. When the great harvest came as predicted, Thales reaped a fantastic profit, and his detractors learned a lesson in humility.

Thales' knowledge of the heavens was legendary by the time of Aristotle; he is even credited by some ancient historians with having compiled a work on navigational astronomy. In his cosmology, he tried to show the same concern for the scientific process as he did in his astronomy and physics. "Thales," wrote Aristotle, "believed the principle of all things to be water. He acquired this notion from observing that not only do things contain moisture but that heat itself is both generated and maintained by moisture."[9] Thales is remembered today not so much for his final observations —for many of his conclusions were incorrect—but for the fact that in all his research, even in art, he used scientific process or argument.

Anaximander (610-546), who lived about the same time as Thales and may have been one of Thales' students or friends, was fundamentally a philosopher, but his metaphysical speculations wound up establishing many of the foundations for later ideas in astronomy, cosmology, and natural evolution. In his cosmology, he talked about a series of fundamental interactions or forces which, by being opposite in their characteristics, created the delicate balance necessary for the creation of all living and non living things. These fundamental forces were born of something he called "The Boundless", a necessary existent in the eternal order. From this idea, later Greek philosophers developed the concept of numerous basic elements that inter-reacted to cause the cosmos. Anaximander also believed that the natural order was not fixed from the start, but designed to allow for continuous change; and the things which made up this natural order, both living and non-living, were so conceived that they would evolve as necessary so as to function regardless of micro or macro cosmic change. He tried to explain his theory of evolution in a book called *Concerning Nature*. Meanwhile, his contributions to technical astronomy included the use of the gnomon for measuring time intervals and the recording of solar changes during the course of the year. Though creative, his scientific ideas had little or no scientific evidence to back them up. For instance, he guessed that the stars and planets were wheel-like constructions of compressed air which blew out fire and heat from slits in their surfaces. He thought the Earth was a mere 5 percent of the Moon's true size and that it was mere optical illusion that gave the stars the appearance of being further from Earth than the Moon (Fig. 1-7). He believed that space itself was infinite and that there is a plurality of forms. He

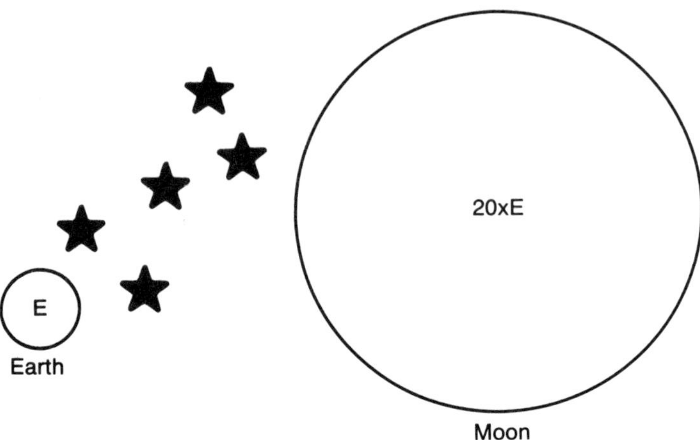


Fig. 1-7. Anaximander had many ideas that were ahead of their time. He had developed his own evolutionary theory which argued that everything in the universe evolved from some ancient form. But Anaximander held that the Moon is twenty times larger than Earth and the stars are much closer to the Earth than it is.

taught that the Sun, the Moon, and the stars resulted from the dispersion of a great sphere of fire that encircled the Earth. Earth itself, he taught, was cylindrical with a depth one-third its breadth and its distance from all other celestial objects was equal from every direction. Man, he believed, evolved from animals and man's history could be traced back to the sea and specifically to the fish that inhabited the sea.

In the years just before Anaximander's death and for twenty years afterward, a new voice in science and philosophy came to national attention. It belonged to Anaximenes (585-526), who had for years been both a student and friend of Anaximander's. He taught that his mentor, and Thales before him, were wrong in what they believed were the basic substances of the universe.[10] Air, not water, filled the universe. This was an idea much more acceptable to people in general, for it was apparent that there was more air than water within and without all bodies. But despite the wide appeal of many of his arguments about the cosmos, many more of his ideas must have left his contemporaries dazed: Earth, Sun, and Moon are flat disk-shaped worlds that float on seas of air; stars do not revolve around the Earth but circle about the top of it; it is not the stars themselves that actually rotate, but the great ceiling and walls of the sky to which the stars are fixed. Plutarch tells us that Anaximenes thought that Earth was formed by the compression of the air about it—a theory not too far from some current cosmological concepts except that Anaximenes never imagined that this air consisted of great clouds of dust and gas.

The next great names in the scientific arena of ancient Greece are those of Pythagoras (570-500) and Xenophanes of Colophon (579-478).

Xenophanes is sometimes credited with having founded the Eleatic school of philosophy, which fostered the notion that there was unity and eternity in all things. He was supposedly exiled from his homeland, and for 67 years he lived as a wandering minstrel, using his poems and essays as a running commentary on social and political events.[11] This is, of course, quite possible; but the chronology and time spans accounted for in this scenario do not fit well into the master puzzle of ancient history. It is, however, true that in the last 21 years of his life, Greece was involved in the Persian Wars begun when the Ionians were led into revolt against the Romans and the Greek tyrants that supported the Romans; Xenophanes may have been forced into exile during this time. Xenophanes would have been in his seventies by 499, so that while he may have been wandering about the Grecian states, it must have been long before the Persian wars. In any event, the chronology of his life is less important to us than his ideas in astronomy. And, as the ancients tell us, Xenophanes taught that the setting of the Sun occurs because the Sun travels at an almost infinite distance from the Earth. Its light burns away during the latter part of the day but is rekindled later on to allow for the coming sunrise (Fig. 1-8). The Moon, he believed, was the source of its own light; and like the Sun, the Moon's light faded with the passing days until finally rekindled after every month. Eclipses, he believed, occurred when the Sun's light burned out ahead of schedule.

Pythagoras was the founder of one of the more famous schools of philosophy in Greece, but it is impossible to separate his original ideas from what the Pythagorean movement taught in the years after his death. What we know of Pythagoras' teachings, we only know through the recorded ideas of his disciples. The man himself is as much a mystery as his true teachings; even Aristotle could only refer to the Pythagorean school and not to the man. "The Pythagoreans," wrote Aristotle in his Metaphysics, "who were the first to take up mathematics were not only the first to advance this study but, having been so fascinated with it, believed that its principles were the principles of all things."[12] As Aristotle points out, to the essential nature of everything there can be ascribed a number—or so the Pythagoreans taught. Just as their predecessors at one time or another said that air or water, or these and other elements, were the basic substances of the universe, the Pythagoreans claimed that it was some mathematical value. All of matter in any part of the universe was fundamentally numerical. Those who would search for the true reality of the world, its constructions, and their relationships, could find their truth only in mathematics. All the world could be defined and described in terms of the four critical numbers: one, two, three, and four. In the Pythagorean scheme, the number one represented a point, the number two a line, the number three a plane, and the number four a solid. As the sum of these critical numbers is ten, the Pythagoreans respected the number ten as the most perfect number.

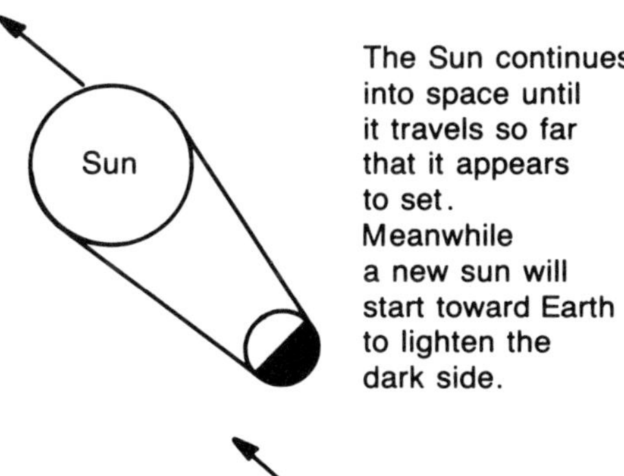


Fig. 1-8. Xenophanes believed that each sunrise was brought about by a brand new Sun.

From this numerical scheme, the Pythagoreans constructed their cosmology and their astronomy. As space was defined as the *unlimited*, it became the central fire about which all the planets and stars revolved. The Pythagoreans fixed all planets and all stars to ten revolving spheres. They decided that the universe must be constructed on these ten spheres because they saw the heavens as being essentially perfect and thereby requiring expression through a perfect number—ten.

Each of the spheres in the Pythagorean scheme represented one of the planets (including the Sun and Moon) or all the stars. Closest to the central fire was a world they called Counter-Earth, second was the Earth itself. Then, in order, came the Moon, the Sun, Mercury, Venus, Mars, Jupiter, Saturn, and finally the fixed stars. Counter-Earth was necessary in the Pythagorean scheme in order to account for eclipses. Since shadows appear on the Sun or Moon during solar or lunar eclipses, the Pythagoreans needed the existence of another celestial object to account for these shadows. It was also the reason why the central fire could not be seen by those on Earth, for Counter-Earth existed between Earth and the central fire.

The Pythagoreans were the first of the ancient schools to reject the idea of geocentricism—that is, the idea that the Earth is the center of the universe. This was a strikingly bold challenge to an accepted idea in ancient cosmology. Only in a nation where science could remain unintimidated by religion could it have evolved, and Greece was just that—a nation where religion and science each had their place and their freedom. But the Pythagoreans not only believed that the Earth was relatively insignificant in the grand scheme of things, they also taught that it was a planet, just like Mars or Venus! However, all of their ideas were not so far-reaching. For instance, they also believed that the intervals between planets coincided with the intervals between the notes on the numerical scale; this concept they called the "music of the spheres." This music, they claimed, could not be detected by the mere mortals because it is a divine music much too subtle for human ears.

Next in line in this chronology of ancient astronomical ideas are those of Anaxagoras (500-420), the philosopher from Clazomenae. He was a man of extraordinary scientific ability and creativity. While most of the Grecian ideas on astronomy before and well beyond his time were simply the results of wild speculations easily challenged by simple observations, Anaxagoras' ideas include the insights that could have advanced astronomy considerably.

Anaxagoras is the man who taught Socrates, and a man who remained very influential in Athenian society until he began to tell his students and his people some astounding things about the sky and the Earth, things which people in his time just could not grasp or tolerate. His almost total disregard for the teaching in astronomy until his time is said to have been the result of his coming across and investigating a meteoric stone at Aegus Potamui sometime around 468 B.C.

Anaxagoras taught that:

- The sun is a red-hot world of stone.
- Earth and the Moon are made of stone.
- The Moon does not give off its own light but reflects the light from the Sun.
- Earth, the Sun, the Moon, and the planets, as well as the stars, have all evolved from natural causes and their evolution began with the setting in motion of infinitely small particles that crashed together to form larger bodies.

He also believed, as Aristotle explained, "that the principles of nature are infinite in number, for he says that most things which are constructed of parts similar to themselves, in the manner of water or fire, are caused and destroyed in this way, only by the aggregation and separation of these parts, and are not in any other sense created or destroyed, but rather remain in existence eternally."[13]

The above represent the more outstanding and valuable of Anaxagoras' insights. But not all of his ideas were so dramatically on target. He also believed that the stars gave off light as a result of their whirling motion, that the Earth is flat and floats on air like a boat floats on water, and that the Earth is indeed the center of the universe.

After Anaxagoras, the next great contributors to astronomy are often mentioned together, as though they were colleagues and companions throughout life. But one of them was teacher to the other, and there was probably a twenty to thirty year difference in their ages. The teacher was Leucippus (fifth century B.C.) and his student was Democritus (460-370 B.C.). They were the founders of Greek atomism, which is the idea that there exists an infinite number of elements all existing in an eternal space. When these elements interact, processes begin which create bodies or affect the order of things. This extends some of Anaxagoras' ideas, although Anaxagoras never developed ideas about atomic existences lying at the core of all creation in quite the same degree that Leucippus and Democritus did.

While little is known about Leucippus, he may have been born in Miletus or possibly Abdera. He may have also been a pupil of Zeno, a Greek philosopher noted for his ideas about motion and infinity.

A bit more is known about Democritus, who spent most of his life in Abdera, located in Thrace. Before settling in Abdera, Democritus evidently travelled widely and studied both philosophy and science extensively.

Democritus was more than a theorist. He wasn't interested in abstractions or purely metaphysical treatments of important subjects. He liked concrete presentations. Even when he reached into the depths of his imagination to offer explanations for certain physical phenomena, he made sure that the models he created from them were logical and defensible. He adopted the views of Leucippus on atomic structures because he was able to expand the idea quite satisfactorily into an explanation of the order and evolution of things.

According to Democritus, all existing things are composed of the same fundamental matter or atoms; these atoms vary in shape, size, weight and possibly in the way they interact. His theory was that the atoms, which he believed to be in perpetual motion, acted and interacted in such a way that they were responsible for the creation of the cosmos—cosmos meaning the *order* rather than the *chaos* in the universe. These eternal and therefore indestructible atoms fall or move at a force relative to their weights, and this constant change in their position as well as the actual motion of the atoms results in a great whirlwind of activity; this, in turn, causes an aggregation of particles so that massive bodies are eventually constructed. This is the way, he explained, that Earth, the planets and the stars were formed, the heavier atoms resulting in the formation of Earth and the lighter ones resulting in the formation of other celestial objects. (It might be noted that one of the current theories about the evolution of planets is that the planets evolved from colliding particles adhering to each other.)

Democitus was still teaching and writing when three other famous Greek philosopher/scientists came into celebrity. These were Plato (427-347) and his famous students, Aristotle (384-322), and Eudoxus (408-355). There was a fourth, who also lived while Democritus still taught, but he was too young to have shared the limelight at the same time: Heraclides (388-315).

Plato was a noble by birth and his choice to travel about in the search of knowledge, to find young disciples, and to found a school of philosophy must have at first been a lifestyle frowned upon by his family, for their first choice of careers for him must have been in politics or trade, or any occupation that would add to family fortune and prestige. His family had provided him with an extensive formal education and he did live for a year or so at the court of Dionysius the Elder (the Greek tyrant also noted for his literary endeavors). One of his teachers was the famous Socrates, a man who had little use for a subject like astronomy and felt that the true concerns for men of philosophy must be man, his institutions, his interpersonal communication, and his politics. Plato was not in agreement—he was as much a scientist as he was a philosopher—yet, no doubt through Socrates' influence, Plato leaned more to metaphysical investigations than to science. Still, he saw philosophy and science as one, an idea that he continued at the Academy, the school he founded and by far the most famous school of learning of the time.

He was a prolific writer, having written more than two dozen dialogues, many epistles, and many Greek anthologies. He was a man of endless dreams and many goals. And while he had accomplished a great deal by the time of his death, there were two great dreams of his which he could not fulfill. The first was to conduct a political experiment in Syracuse which would lead to the foundations of the perfect state. The second was a trip to northwestern Persia, now Iran, to a country called Media, where lived the Magi; from this Median tribe, he hoped to bring back to Greece a whole new source of astrological, astronomical and metaphysical knowledge.

Plato believed that the heavens were examples of divine perfection and were meant to be lessons from the gods on how to learn and live, and by observing the harmony apparent in celestial interrelationships man could find the basis and reasoning for developing harmony in his own social and political life. But there was also practical advantage in studying the heavens, for by the planets and stars man could learn to measure time with great accuracy and understand ever more the agriculture and navigation of the Earth; additionally, by studying the movements of the heavens, man could learn the proper methods for scientific investigation. This last seemed to Plato to be the most important advantage of astronomy. It was not the specific information gathered from observation that was of prime importance in astronomy, Plato argued, but rather the inspiration that came once celestial phenomena could be understood. A study of the heavens could help a man or woman understand the true nature of both sensual and spiritual entities. He would have made the heavens the most important school and his Academy the second, for in the heavens, the lessons came

**Table 1-6. The Order of the Planets According to Plato and Ptolemy.**

| Plato | Ptolemy | Actual |
|---|---|---|
| Moon | Earth | Sun |
| Sun | Moon | Mercury |
| Venus | Mercury | Venus |
| Mercury | Venus | Earth |
| Mars | Sun | Mars |
| Jupiter | Mars | Jupiter |
| Saturn | Jupiter | Saturn |
| | Saturn | Uranus |
| | | Neptune |
| | | Pluto |

from God, and Plato believed in one God who created the heavens and the Earth. Still, a study of the heavens could not be a study of the real world in Plato's scheme of things, for he believed that the world in which we live is not the real world but only a manifestation of true reality. The real world, he believed, must be pure, eternal and perfect, for it was created by God and any creation by a perfect being must necessarily be as perfect as He.

To Plato, Earth was round and inorganic, but the stars were living things existing in eternity, spherical bodies of fire that inhabit a universe which is also a living thing, formed to perfection in the shape of a globe. All movement in the universe is caused by soul, for Plato believed that only soul could originate motion. In Plato's order of the universe, there were seven wandering stars (the Moon, the Sun, Venus, Mercury, Mars, Jupiter, and Saturn) as well as the fixed stars which have their stations in the outer limits of the universe. He believed Earth is stationary and mistook Venus to be closer to the Sun than Mercury. In Book 10 of *The Republic*, he gives his ideas about the relationships of the stars and planets by telling the story of a soldier, Er, who was killed in battle, lay dead for twelve days and then, as he lay on his funeral pyre, came back to life. To the startling witnesses of his return from the dead, he told the story of his soul's journey during those twelve days and explained how the universe looks to souls in heaven. Er described a pillar of light which connects all the heavens and something called the Spindle of Necessity, by which the revolution of heavenly bodies is maintained. There are eight whorls in this cosmological scheme, fitting like bowls of different size, one into the other, the rims of these bowls having different widths and different colors.

> *. . . The rim of the greatest whorl exhibits a variety of colours; that of the seventh is most brilliant; that of the eighth derives its colour from the reflected light of the seventh; that of the second and that of the fifth are similar, but are of deeper colour than the others; the third has the palest colour; the fourth is rather red; and the sixth is almost as pale as the third. Now the distaff as a whole spins round with uniform velocity; but while the whole revolves, the seven inner circles travel slowly round in the opposite direction; and of them the eighth moves quickest, and after it the seventh, sixth and fifth, which revolve together; the fourth, as it appeared to them, completes its revolution with a velocity inferior to the last mentioned; the third ranks fourth in speed; and the second, fifth. The distaff spins round upon the knees of Necessity. Upon each of its circles stands a siren, who travels round with the circle, uttering one note in one tone; and from all the eight notes there results a single harmony. At equal distances around sit three other personages, each on a throne. These are the daughters of Necessity . . . who, clothed in white robes, with garlands on their heads, chant to the music of the sirens . . . of the past . . . of the present . . . of the future.*[14]

As Plato relates Er's story, it is clear that it has been contrived so as to agree with traditional ideas such as the music of the spheres, the then-known planets, and the Greek belief in the immortality of the soul. Plato's use of eight revolving spheres was also no doubt based on the Pythagorean idea of ten harmonious spheres. All in all, the scheme is fantastically precise, although hardly a true representation of the order of the universe; it can be appreciated only as a metaphorical presentation. Plato seemed to enjoy dealing in metaphors, particularly when it came to discussing astronomy. But then ideas were always more important to Plato than facts. To Plato, there were just too many illusions, too many uncertainties in science for anyone to be sure of anything at all. What we perceive as true facts may not be true at all, for our world of sensual things presents everything in distorted fashion, much as the reflection of light in water leaves us to expect a stick held half below the surface is bent out of shape.

Aristotle avoided all dependence on metaphor or abstraction and preferred to try and express his theories in the form of concrete models. Born in Sagira, he developed a keen interest in medicine from his father, who was a physician. His interests in philosophy and mathematics came as a result of Plato's influence under which he studied for nearly nineteen years beginning in 366 B.C. He tutored Alexander the Great and founded a school at Lyceum. He taught that the basic components of nature were earth, air, fire, and water. Many of his ideas on astronomy can be found in his *On The Heavens*, where he expresses that the heavens always existed, never came into being, and therefore can never be destroyed, for that which has had no beginning can have no end. "There is no beginning or end to its duration," he wrote, "for it contains and embraces in itself the infinity of time."[15] He added that the shape of the heavens must be spherical for "that is the shape which is most appropriate to its very substance as well as that which is required by nature primary."[16] The heat and light which the stars generate, he explained, "are brought about by the function caused by their motion through the air."[17] This last Aristotle concluded by observing missiles which undoubtedly caused force as they moved. As missiles move, he explained, they themselves are "fired so strongly that leaden balls are melted; and if they are fired in such a way, the surrounding air must also be so affected. Now, as the missiles are heated because they move rapidly through the air, which is converted into fire by the rapid movement, the (stars') upper sections are carried on a moving sphere so they themselves are not caught in flames though the air underneath the body's revolving sphere is necessarily heated by the motion, and particularly in the part where the sun is shining on it. Hence the warmth increases as the sun gets nearer or higher overhead."[18]

The stars, Aristotle argued, though spherical in shape, neither spin nor roll as astronomers sometimes suggested; as a matter of fact, he stated, they have no movement at all.[19] "And the theory that the movement of the stars produces harmony is untrue."[20] Earth, he went on to explain, is also spherical but like the stars it has no spinning or rolling motion. Aristotle believed that on the cosmic scale, Earth is a rather insignificant object.

In Book 3, Chapter 14 of *On The Heavens,* he wrote:

> *Again, our observations of the stars make it evident, not only that the Earth is circular, but also that it is a circle of no great size. For quite a small change of position to south or north causes a manifest alteration on the horizon. There is much change, I mean, in the stars which are overhead, and the stars seen are different, as one moves northward or southward. Indeed, there are some stars seen in Egypt and in the neighborhood of Cyprus which are not seen in the northerly regions; and stars, which in the north are never beyond the range of these observations, in those regions rise and set. All of which goes to show not only that the earth is circular in shape, but also that it is a sphere of no great size; for otherwise the effect of so slight a change of place would not be so quickly apparent. Hence one should not be too sure of the incredibility of the view of those who conceive that there is continuity between the parts about the pillars of Hercules and the parts about India, and that in this way the ocean is*

*one. As further evidence in favor of this they quote the case of elephants, a species occurring in each of these extreme regions, suggesting that the common characteristic of these extremes is explained by their continuity. Also, those mathematicians who try to calculate the size of the Earth's circumference arrive at the figure 400,000 stades. This indicates not only that the Earth's mass is spherical in shape, but also that as compared with the stars it is not of great size.*[21]

Aristotle argued that the Sun, Moon and planets consisted of an additional necessary substance which he defined as the ether, and which was the source of scientific inquiry up until the latter part of the nineteenth century. The ether, according to Aristotle, extends from the heavens and toward the earth but no further than the Moon. Below the Moon, the ether gives us the four necessary substances of Earth: earth, fire, air and water, with air and fire being more closely associated than the other two. As for the specific arrangement of the stars and planets according to Aristotle's scheme, it is better to tackle this subject after a brief introduction to the geometry of space as it was conceived by another of Plato's pupils whom Aristotle knew well.

His name was Eudoxus of Cnidus and he lived from about 408 to 355 B.C. Besides having studied under Plato in Athens, he may have done additional studies under Egyptian tutors in Heliopolis, Egypt before he finally settled in Cnidus where, according to ancient writers, he constructed his own observatory. Now, an observatory in those days would qualify today as little more than a study containing collections of astronomical works, sky charts, astronomical instruments mostly used to measure time, and perhaps a building designed to function as a celestial detector—that is, it would be so designed as to capture light from the Sun or Moon on certain days of the year or allow the formation of shadows which would signal some astronomical event.

Eudoxus is credited with having determined the length of the solar year and with having developed a calendar quite similar to the Julian calendar which remained in use for centuries. He was, perhaps, the first of the Greek astronomers to explain the position and movement of the planets in a scientific way. Eudoxus designed a theoretical model of the universe which employed numerous concentric spheres supporting the planets in their positions, an idea of such scientific weight that it was still accepted by many astronomers as late as the seventeenth century. His model was really an extension of Plato's idea of celestial spheres with a common axis. But whereas Eudoxus' spheres were simply geometric models for which there was no physical representation in the real universe, Aristotle grabbed the idea and expanded it to include 55 concentric spheres, each sphere having a physical form similar to shells of crystal and each moving about the other. All of these spheres were connected in such a way that the motivating force, generated by the most exterior sphere, was carried within to each successive sphere.

This model of Aristotle's was far too complicated but very impressive in terms of its precision. The model required numerous spheres for each planet and a complex system of relationships for each movement along the spheres, and for varying speeds and distances. Few astronomers today would dare to try and devise such a scheme without the help of a computer. How did Aristotle find time to sketch it out while writing, teaching, and researching everything from the physics of light to the idea of animal soul?

Both Plato and Aristotle were prolific writers. They were men who traveled greatly, spent much of their time lecturing and teaching others as well as mixing socially and politically—yet they somehow found the time to produce volumes of words. Today a writer scrawls notes on a piece of paper with pen or pencil and then types his manuscripts at 40 to 60 words per minute; he can easily produce and publish hundreds of thousands of words a year. Writing was not so simple in the time of Plato and Aristotle. There were no typewriting machines or even convenient methods of producing graphics or symbols at anywhere near the rate they can be drawn today. Yet they produced so much written material!

How did they do it? Perhaps they employed their students in this manner: As either of them lectured they would assign students to write out sentences. There might be five students, each directed to write out every fifth sentence. As Plato or Aristotle spoke, they need not stop to allow each student to catch up; they needed only to look at the particular student responsible for a given sentence. Later another student would have the responsibility for bringing together the scribbling of the other five students. Did Plato and Aristotle edit their work and rewrite as necessary? They must have. There must have been occasional errors in the transcriptions or thoughts they wanted to change.

Heraclides, the third student of Plato's discussed here, is mainly remembered for his arguments that Earth rotates and that the planets orbit the Sun. But like most astronomers of his time, he was still tied too tightly to traditional ideas about Earth as the center of the universe to challenge this idea. He also mistakenly taught that the stars and the planets move in what is referred to as concentric and eccentric revolving spheres. These spheres, of course, had to be circular, for even the crudest sketch of planetary movements would make that obvious. The largest of the circles in Heraclides' scheme was called the deferent, and the smaller was called the epicycle. The planets moved in the epicycle while the center of the epicycle moved along the circumference of the deferent. Any movement occurring along the epicycle or along the deferent must be uniform. The theory worked fine except for two wayward planets that would not adhere to the phenomena, Mercury and Venus. Aristotle offered his aid in trying to resolve the theoretical problem occurring in Heraclides' scheme and in so doing revised the model so that the spherical theory applied to all moveable celestial objects *except* Mercury and Venus. It seemed to work fine on paper but somehow Aristotle was never satisfied with it. Heraclides was also dissatisfied with Aristotle's scheme and decided that he could "save the phenomena" by having Mercury and Venus orbit the Sun.

At the same time that the Greeks were devising new cosmological schemes and adding to their knowledge of the skies, a Babylonian priest, who lived in the early part of the third century B.C., had been researching ancient texts and coming up with some startling information about the age of the universe. Additionally, he was doing some observations of the skies on his own, as well as testing astrological theories.

His name was Berosus (Bel-Usur). Originally from the city of Bel Norduck, he decided to leave his homeland when Alexander the Great sent his forces against the city. During his travels he studied Egyptology, completed a history of the Chaldeans, taught astrology and astronomy, and made some efforts to contribute to the Greek medical arts. In medicine, however, he must have been on awkward footing. His deep involvement with mythology, astrology, and the occult must have put him on the wrong side of the disciples of Hippocrates.

Little of the scholar priest's background or writings is known, except for what is revealed in the works of ancient writers such as the Jewish historian Flavius Josephus (first century A.D.) and the Greek historian Eusibius Pamphili (263-339). Berosus is mainly remembered for passing on the history of Babylon to the Greeks. But he is also credited with having developed a type of sundial called the hemicycle which was a variation of standard timing instruments of his time. And he is recorded as having taught that the Moon is a sphere, half of which gives off its own light, and that Earth is much older than anyone imagined, perhaps as much as a half-million years old. He arrived at this time span after poring through many written creation myths, and recording many oral histories of the Mesopotamian civilizations.

Shortly after the time of Berosus, there lived a Greek scholar and astronomer who added greatly to Greek astronomy, though his ideas were a lot stranger than the ancient myths to his contemporaries and, unfortunately, were left to collect dust. His name was Aristarchus of Samos (320-250). Historians have his ideas secondhand, for his works evidently went up in flames during the final destruction of the great university and celebrated libraries in Alexandria in the fourth century A.D. The only one of his works which scholars have actually had access to is a treatise which estimates the sizes and distances of the Sun and the Moon. Aristarchus organized his treatise as a series of geometrical propositions, a popular format for scientific writing in his time. If anyone in those ancient times could have carried astronomy from the theater of art to the laboratory of science, it was Aristarchus. He was able to prove that the Moon shines with light reflected from the Sun, that the Sun is more than 18 times more distant from Earth than the Moon, and that the Sun is from six to eight times larger than the Earth. His ratios were hardly accurate, but they represented giant steps in the right direction. His problem was his reliance on a geometric investigation of planetary characteristics; this meant that the specified quantities would always be described in terms of ranges rather than exact values, as in "from six to eight times," or "18 times but less than 20 times."

Aristarchus also proved another spectacular fact but his argument was finally only appreciated for its mathematical efficiency than its startling revelation, for the Greeks were men of numbers and their application before they were men of the stars and planets and their interactions. The Sun, he tried to show his countrymen, was the center of Earth's orbit. His was the first argument for a heliocentric world structure but as his ideas did not coincide either with observational evidence of the time or the weight of historical arguments, the idea of a Sun-centered solar system was put to rest for about seventeen centuries. Despite his great breakthroughs, Aristarchus never challenged the idea of the stars being fixed in space.

Eratosthenes (275-195), a contemporary of Aristarchus for a short time, is the next great name in the history of Greek astronomy. He was actually a geographer, but he also wrote on mathematics, astronomy, and philosophy. He is famous particularly for having come up with a fairly accurate measure of Earth. Using points of reference at two different cities, Alexandria and Syene, he took the angle of the Sun's elevation from the horizon on the longest day of the year. He knew almost the exact distance between the selected positions within each city, about 496 miles. He also knew that on the day he selected, the Sun was positioned 90 degrees from the equinoxes, the two points on the (imaginary) celestial sphere where the ecliptic (the circle on the celestial sphere that is located on the plane of the Earth's orbit) and the equator meet.

It was apparent to Eratosthenes that the noontime Sun over Alexandria was not exactly overhead at the time of measurement. But it should have been if it were indeed noontime, for certainly in Syene it was directly overhead at noontime. Actually, Eratosthenes never expected the Sun to be directly overhead at noontime in both cities at the same time, although less knowledgeable citizens of both cities might have expected such a contradiction. Determining that the Sun was actually 7½ degrees from its expected overhead position in Alexandria, Eratosthenes now knew that the difference in latitude from Syene to Alexandria was also 7½ degrees. Now, all he needed to do was find the multiplier for the final computation. This he found by dividing 7½ into the number of degrees in a circle. The result is roughly 48. Now he multiplied 48 times the 496 miles between the two cities and arrived at a figure of 23,808 miles. He wasn't off by much, for the circumference of the Earth is 24,902 miles.

At the same time that he resolved what the size of the planet Earth must be, he also confirmed that it was round. But all this came across as just too much fiction—just another scientist trying to attract attention. A disbelieving public could not accept that Earth was indeed so large nor could they fathom that they were living on a round world. Thus, Eratosthenes' ideas were filed away with those of Aristharcus. The world would have to wait for Christopher Columbus for confirmation that the world was indeed round, although many astronomers, navigators, and adventurers already knew the truth long before Columbus. Undaunted by the great criticism he had to field, Eratosthenes went on to measure the Moon and the Sun and their probable distance from Earth. Here his measurements were not nearly so accurate, but they were much closer to the truth than anyone else's up until his time.

As we approach the last four centuries of the late ancient period, there are only four discoverers left to meet. They are Hipparchus (190-120), who was in his time what Tyco Brahe was in his (that is, the keenest observer of the heavens and the most accurate recorder of his observations); Posidonius (135-57), who added greatly to knowledge about the atmosphere and also produced remarkably accurate measurements of Earth; Lucretius (98-55), the hard-living poet whose ideas on the formation of celestial bodies foreshadowed current theories of galactic birth and evolution; and, finally, Ptolemy (100-170) who synthesized, verified, and expanded the ideas of earlier astronomers, particularly those of Hipparchus (Fig. 1-9).

Hipparchus came from a country in the northwestern part of Asia Minor called Bethynia, which, if it were still in existence, would be located within the political boundaries of what is now Turkey. Scholars guess that he was born about the year 190 and died around 120, but there are no dependable references in early histories to make these dates certain. Appreciated during his time and for centuries afterward as the most careful and scientific observer of ancient times, he remained convinced that the heliocentric model of the universe conceived by Aristharcus was wrong and that the Earth was both motionless and the center of the universe. If he had only verified the ideas of Aristharcus, astronomy throughout the rest of the ancient period and throughout the middle ages would have a far different history; but as observational astronomy had to wait for the telescope before it could enter its age of discovery, the effect acceptance of the heliocentric model would have had over the long run (that is, to modern times) is debatable.

Among the many things for which Hipparchus is remembered in astronomy are his measurements of star positions, his estimation of the tropical year, his star catalog, and his relative measurements of Earth, the Sun, and the Moon and their distances. He is probably best remembered in astronomy for having realized something called the "precession of the equinoxes," or what is sometimes referred to as the "precession." His discovery came about because he had had access to Chaldean tables that noted some measure of erratic precession; additionally, there was a marked difference in the positions of the stars he was observing from what was recorded as their positions by astronomers of past centuries. This led him to assume that the equinoctial points must change with each passing year. This needs some clarification.

To understand what is meant by the precession of the equinoxes, remember the definition of an equinox: An equinox is either of two points on the celestial sphere where the ecliptic and the celestial equator intersect. The ecliptic is the circle in the plane of Earth's orbit and the celestial equator is a circle representing Earth's equator. You will see that there are indeed two points at which the ecliptic and the equator intersect. The equinox (point) at which the Sun crosses the celestial equator as it moves northward is the *vernal* equinox. The equinox at which the Sun crosses as it moves southward is called the *autumnal* equinox. As it turns out, these solar crossings occur twice each year, usually on March 21 and September 23. The dates mark the spring and autumn in the Northern Hemisphere and autumn and spring in the Southern Hemisphere. When the Sun makes either crossing, we experience a day when daylight and nighttime are each 12 hours in length.

Hipparchus noted that there was a westward motion of these crossing points along the ecliptic. Astronomers now know that this westward precession is the result of the gravitational interaction between the Moon, the Sun, and Earth. The result of this gravitational field also forces the celestial poles

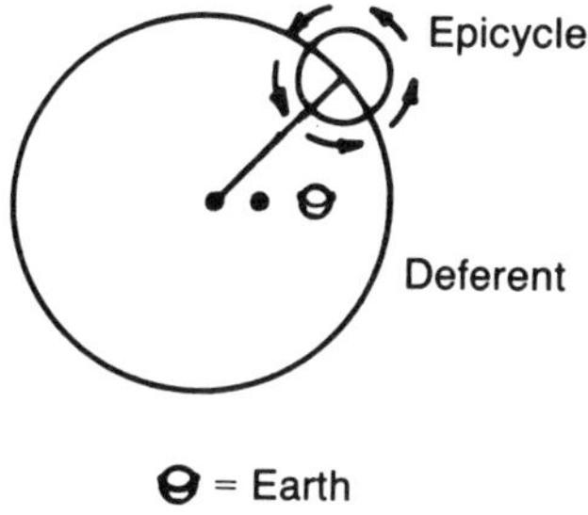


Fig. 1-9. In Ptolemic astronomy, the planets move along a circle (called the epicycle) which has its center coincident with the circumference of a larger circle (called the deferent). The epicycle moves about the larger circle, its center always on the circumference.

into circular movements on the celestial sphere. The result is that during different periods, different stars are noted at relative positions to these poles. It is calculated that the precession comes full circle every twenty-six thousand years.

Once he realized that the equinoxes moved along the ecliptic and he arrived at an expected value (roughly 45″ of arc) to represent the annual changes in star longitudes, he was able to make a relatively close estimation of the time it would take the Sun to return to a given equinox (called a tropical year) and of the time it would take the Sun to return to a given fixed star (sidereal year).

Hipparchus' star catalog was completed about 129 B.C. and listed one thousand stars. In the catalog, he not only listed star positions but also gave the apparent brightness for each entry, using a system of magnitude specification on which modern systems are still based. His was the first star catalog to be produced, and it was so accurate that it proved to be of invaluable aid to Ptolemy some 250 years later, and was even a source of reference for astronomers right up until the eighteenth century.

In his studies of the Sun, Hipparchus drew heavily from the observational data compiled by his contemporaries and those astronomers who lived before him. Some of his observations received a great deal of criticism and were eventually revised by other astronomers. Some of his observations resulted in some unusual theoretical comparisons. For instance, he was able to show that the motion of the Sun, Moon, and planets were equally explained by two separate cosmological schemes, that based on the idea of orbit about a deviating center and that based on the complicated system of epicycles and deferences, wherein the circle representing the orbital path of a planet has a center which moves in another orbit represented by the circumference of a larger circle. He also explained that the length of the year depends on the return of the same equinoxes and solstices that determine the seasons.[22] This observation was the first of a series that led him into his discovery of the precession of the equinoxes, for he went on to arrive at the true length of the year to be 1/300 of a day less than 365¼ days. Realizing that this figure was less than that represented in Babylonian calculations, he felt that the only way to explain the contradiction was to assume that the interval between returns to equinoctial points and the interval between returns to the same star were not the same.[23]

In his studies of the Moon, Hipparchus was always sure that there was some mystery to the constant irregularities of lunar motions. He had done some ingenious studies in which he applied the epicycle theory to account for some of these motions, but there were forces involved here about which Hipparchus could not possibly know a thing about as the Moon's motion is characterized by many perturbations resulting from gravitational influence by the Sun and Earth.

Just as he was disturbed about the unsolved mysteries of lunar motions, he was also disappointed in his planetary observations and research. He would have liked to have been able to devise a mathematically sound model of planetary motion but he found the task too impossible, for he noted that, as in the case of the Moon, there were also too many inequalities in planetary behavior and none of the theories of his time, when examined individually or together, could successfully explain the inconsistencies. He believed a new cosmology was necessary, and he personally felt that what was needed was some synthesis or expansion of current theories.

Posidonius (135-50) was another astronomer noted for his keen observations and adherence to scientific process. He, too, compiled many accurate tables, was respected by his contemporaries and was the source for future astronomers, but his reputation never quite came to equal that of Hipparchus'. He is remembered in history mainly as a stoic philosopher who carried on the teachings of Plato and Aristotle. In the field of astronomy, however, he is remembered as having obtained some of the best estimations of the Sun's distance from Earth and the Sun's diameter. He also made an attempt to measure the distances and diameters of the stars. His system of measurement was based somewhat on that used by Eratosthenes to measure the circumference of the Earth. For instance, to determine the distance to Canopius, one of the brightest stars in the sky, he noted its position relative to points at

Rhodes and Alexandria. At the time he took his measurements, Canopius was seen from Rhodes to be slightly above the horizon when looking south, but in Alexandria, the star was at a significantly higher altitude. Posidonius estimated, then, that the distance of the star over the horizon was roughly 500 stades and its circumference roughly 240,000 stades. He was, of course, way off in judging the size of the star and also in judging its true distance from Earth, but his efforts were commendable.

To this point, we have considered mainly the Greek astronomers. Rome, too, had its sky-watchers, but the only Roman to have any valued impact in astronomy, though indeed his contributions were mainly in the area of cosmology and he was a poet rather than a scientist, was Lucretius Caris (first century B.C.). Little is known of his personal life or character except what a few ancient historians have occasionally noted and what his poem, *On the Nature of Things*, reveals about him. The poem is particularly important because it carries forth the atomic theory of Epicurus, the Greek philosopher and moralist, to the Roman people at the same time that it challenged a Roman religion covered in superstition and fear of the wrath of the gods. In Roman mythology, the gods play continuously at manipulating the world and the destinies of men; Lucretius challenged the whole idea and wrote that if the gods did exist, they could have little concern with the offices of men and women, for as they were immortal or eternal, nothing men or women could do would have any effect on them.

Lucretius pictured the universe as an infinite void consisting of invisible particles of matter called atoms. (The word "void" in ancient times had a slightly different meaning than it does today; it referred to something needing purpose and order and consisting of nothing recognizable or valuable.) These atoms Lucretius believed to be eternal, indivisible, and indestructible, yet they were made up of sections, each section having a specific part to play in the life of the atom; but these sections could not be separated from each other. Today, of course, we are familiar with the fact that physicists have been able to "split the atom." But splitting the atom is really a contradiction in terms. What physicists today call an atom is not what Lucretius actually had in mind. He referred to a basic unit of matter (called later on in philosophy a "monad.") In Lucretius' scheme of things, all "atoms" would have remained distinct existences if it were not for some phenomena that caused them to begin swirling rapidly around each other instead of continually falling through space. The swirling motion caused collisions and adhesions; these adhesions continued until larger and larger objects were formed. Eventually worlds the size of Earth and larger were formed.

All the important work and traditional theories that stood the test of time from Thales' generation through the mid-second century B.C. were synthesized, evaluated, and copied by a man who was to have a profound effect on astronomy for more than a thousand years. His name was Ptolemy and he lived from 100 A.D. to 170 A.D. His influence came from his exhaustive accumulation of factual evidence, his mathematical formulas and applications, and his sound arguments. Unfortunately what he passed on to future generations actually inhibited rather than enabled the continuing development of astronomy as a science. Ptolemy was primarily a follower and a strong advocate of the great technician, Hipparchus, and he therefore accepted, seemingly verified and fully supported a geocentric model of the universe.

Probably born in Ptolemais Hermuii, Ptolemy made most of his contributions to astronomy in the second quarter of the second century. It is his work which summed up the efforts of astronomers before him for the centuries of researchers and observers that would follow. His most famous text, originally titled *The Mathematical Collection*, then *The Great Astronomer,* has come to be known by its Arabian title: *Almagest*.

The *Almagest* served as an encyclopedia of astronomical data. It consists of 13 books dealing generally with the following topics: Earth and the spherical movements of the heavens; the Sun, the Moon, and the "fixed" stars; and Mercury, Venus, Mars, Jupiter, and Saturn and their heliacal risings and settings. Great scientist that he was, Ptolemy never forgot the theological implications that affected any search for understanding of the heavens and he notes this in the preface to Book 1: "For given that all beings have their existence in matter, form and motion, yet none of these can actually be seen, but only conceived, in its subject matter separately from the others, if one should seek out in its simplicity the first cause of the first movement of the universe, he would find God, invisible and unchanging."

To have compiled *Almagest,* Ptolemy must have researched the ideas of all the Greek astronomers surveyed in this text as well as to have gone over the findings of the lesser known ancients. He would also have been aware of the fifth century work of Metan, who introduced the Metanic cycle, a cycle of 19 years which includes 235 lunar months, and which to this day is used to establish the dates for Easter, as the moon's phases occur on exactly the same dates from the beginning of one Metanic cycle to the other. And he would have been just as familiar with the work of Callippus of Cyzicus, the fourth century B.C. astronomer who established the Callippic cycle (a modification of the Metanic cycle) and who also developed a system which explained solar and planetary interactions. Callippus' work was especially important, for in his time it was believed that the true month was 29½ days long; and the result was that the ancients had fixed the calendar so that a cycle of 235 months (125 months having 30 days and 110 having 29 days) was the equivalent of 19 years. But Callippus realized the equation was wrong, for the embarrassing result was one day too many. Thus, he quadrupled the 19-year cycle and in the new one of 76 years, he reduced one of the 30 day months to 29 days and established the actual length of the year to be 365¼ days. By Ptolemy's time, however, further revision was made to the Callippic cycle, for, as we have already noted, the year is not exactly 365½ days.

What Ptolemy verified about the motions of the Sun, Moon and planets was simply an extension of the work of Hipparchus. His main objective was to verify his geocentric theory, which has since come to be known as the Ptolemic system. Not only was the Earth the center of the universe as far as Ptolemy was concerned, but it was immovable. If Earth did move, he argued, then we would not be able to observe falling objects striking the center of Earth. Additionally, if Earth rotated as some suggested, then some object thrown upwards in a straight line would not fall back to the same spot from which it was hurled. His arguments seem rather childish to us today, but they were forceful enough to affect Arabian and Christian astronomy through the fifteenth century.

In his cosmological order, Ptolemy established the following order for the Sun, Moon, and planets. Earth was first at the center, then came the Moon, Mercury, Venus, the Sun, Mars, Jupiter, and Saturn. Even in Ptolemy's time, Earth was not considered a planet and still only five other planets were known to astronomers. Ptolemy's cosmological system would prevail, though it would occasionally be challenged by astronomers (who, though dissatisfied with it, could not come up with a better scheme) until Nicholas Copernicus rocked the scientific (as well as the religious) community with his heliocentric model.

While Ptolemic astronomy is interesting in terms of its scientific process, its impressive mathematics and its detailed and orderly presentation of evidence, it, nonetheless, is little more than a synthesis of ancient ideas that finally would not stand the test of time. Thus a review of its ideas is not now warranted and we might move now to the middle ages.

## THE MIDDLE AGES

Political boundaries were constantly changing during the ancient period and also in the Middle Ages. Kingdoms would grow into empires and war with other empires. Power and greed was the great motivating factor in ancient times. The Greek city-states were typical examples. While Eudoxus, Plato, Aristotle, and other noted astronomers up until the time of Ptolemy watched and studied the heavens, the Greek city-states were engaged in countless wars and conflicts which gradually tore away at their glory. Greek civilization entered its greatest period shortly after the war with Persia which finally ended in 449 B.C. Athens was the city-state that led the way into a period of sustained cultural genius, one

Table 1-7. Medieval History of Astronomy.

| Name | Comment |
|---|---|
| Arayabhata (fifth century) | Used algebra in astronomy; believed there are mathematical formulas to explain all celestial phenomena; dissatisfied with Ptolemic ideas about planetary movement; believed the Earth is in motion. |
| Brahmagupta (590-660) | Supported most of Arayabhata's theories and also sought to combine the sciences of astronomy and mathematics. |
| Mankah (eighth century) | Hindu teacher of Al-Mansur, the second Abbasid caliph; introduced the caliph to the Siddhanta and whetted the Arabian appetite for the science and astronomy of other cultures. |
| Al-Kwarizmi (780-850) | Verified and updated Ptolemy's *Almagest*, drew up the first star tables in Arabic using the Indian system of counting. |
| Al-Battani (850-929) | Modified some of Ptolemy's calculations, including the true length of the year and the constant of precession. |
| Al-Tusi (1201-1279) | Challenged Ptolemic cosmology but could not come up with a sufficient alternative. |

that left for all time its mark on the arts and sciences. Athens, as students of history will remember, eventually fell during the Peloponnesian War and gave up its leadership to the militaristic Spartans; but the poetry, literature, philosophy, and science that it had nurtured lived on. The glory that was all of Greece's lasted only until the death of Alexander, after which the city-states were torn apart. There were efforts to reunite them but before this could happen the Greek's fell under Macedonian rule and by the middle of the second century B.C. were so weakened by continuous wars that they came under the domain of the Roman Empire. Fortunately, the Romans never tried to squelch Greek culture; actually, just the opposite was true. The Romans showed great consideration, fascination, and respect for Grecian intellectualism.

Rome was, up until the middle of the third century B.C., mainly a continental power, governing with force territories in central and southern Italy. Further expansion of the empire was inevitable, for in those times, and through most of the Middle Ages, any empire that did not seek to conquer could only expect to be conquered. Rome had its great military machine, the resources of a rich land, and ambitious leaders. But, then, so did Carthage. The inevitable result was the Punic Wars from which Italy wrestled away Corsica, Sardinia, Sicily, and Spain. Continued expansion over the next century led to the defeat of the Macedonians and dominion over Greece. Then followed the brief age of the Caesars, more conquests, more internal strife, and more revolt. Rome, however prevailed until about 190 A.D., and with it the Greek culture; then the empire began a rapid decline. In 265, the Persian siege threw the empire in anarchy and under the emperor Diocletian in 283, the empire was divided into four political governments, two in the west, two in the east. Rome itself still endured, though its glorious age was behind it, until the fourth century A.D. when the empire was permanently divided into eastern and western divisions, with the eastern half known as the Byzantine Empire of the East Roman Empire. (The name Byzantine came from a city rebuilt by Constantine I and renamed Constantinople, a name it carried until the fifteenth century.) The Western Roman Empire was short-lived; it quickly deteriorated into anarchy while its territories were under continuous attack; by the end of the reign of Romulus Augustubus in 476, the Western Empire was only a part of history.

With the division of the Roman empire into eastern and western empires, Greece became an integral part of the Byzantine empire, which consisted not only of Greece but also of Epicus, Macedonia, Thrace, and Illyra. The Greek language became the most common throughout the empire, although Latin was the language used in legal and administrative communications. Greek culture and science was still in dominance, but in the fourth through sixth centuries the Hellenistic culture began to wane under Oriental and Arabian influences. In the seventh century, there was a brief rebirth in Hellenistic traditions but then came a smothering by Oriental cultures.

To the beginning of the fifth century, Grecian science was preserved in the Eastern and Western Roman empires. With the fall of the Western Empire, the Byzantine and Arabian cultures fell heir to the science and literature of the Greeks. It is to the Arabians that the great debt is owed for preserving ancient Greek astronomy, for they became obsessed with Ptolemic astronomy and spent centuries studying it and attempting to verify it.

The passage of Grecian ideas to the Arabians was no direct effort, however. Since the late ancient period, and particularly from about the time of the Alexandrian conquests, Greek science had made its way to India. Indian astronomers were fascinated with Greek astronomy and particularly with the scientific approach the Greeks made necessary. But the Indian interest was mainly in the mathematics employed by the Greek astronomers, for the Indians were always fascinated with math. They were also less concerned with purely observational data than they were with the underlying principles that governed the movement of the planets, the Sun, and the Moon.

The really productive age in Indian astronomy, however, was long after the Greeks became a part of the Byzantine empire, perhaps from the middle of the third century through the seventh century, for it was during this period that India was under the rule of the Gupta dynasty and the Harsha culture, periods when the Hindu culture experienced its own golden age. During this time the Indian astronomers Aryabhata and Brahmagupta lived.

Aryabhata lived in the fifth century. A mathematician, he was one of the first to use algebra in astronomy. The works he left behind, included as part of a traditional compilation of mathematical and astronomical writings collectively known as the Siddhantas, include arithmetical formulas, trigonometric measurements and quadratic equations. Aryabhata believed that there were algebraic formulas and geometric principles which could account for all celestial mechanics. Awed as he was with the Ptolemic process for accounting for and verifying astronomical facts, he was never completely satisfied with Ptolemic ideas about the ways in which the planets moved, nor was he satisfied with many of Ptolemy's cosmological ideas. Aryabhata objected particularly to the idea that Earth was at rest; and he was sure enough of his own calculations and observations to argue that Earth must rotate, whether or not it is fixed in one spatial coordinate.

Brahmagupta (590-660) was also a mathematician and astronomer. He wrote a poem called the "Brahma-Sphuta-Siddhanta" (improved system of Brahma), a work on astronomy which contained, also, chapters on mathematics. Well-studied in both the ideas of Ptolemy and Aryabhatta, he preferred to support the latter's planetary theories, as he, too, felt that there was sufficient evidence to prove that the Earth rotated.

In the year 773, Al-Mansur,[24] who was the second Abbasid caliph (from 754-775), received a learned Hindu by the name of Mankah who was able to speak about the heavens as though he himself had been privileged to receive a personal tour. Thoroughly impressed with this learned astronomer, Al-Mansur soon began to realize that the Mohammedan world might be shutting out the light of knowledge by disregarding the accumulated knowledge of other civilizations. When Mankah presented Al-Mansur with a copy of the *Siddhanta* to show the source of his great knowledge, Al-Mansur commissioned Persian astronomers to begin translating the *Siddhanta*. This was just the beginning! Their taste for this new-found knowledge of the skies whetted, the Arabians began translating even Greek texts, aware that they were the source for much of what the *Siddhanta* contained. Among those texts acquired in the early ninth century by the Arabians was Ptolemy's "Very Great Composition," his great astronomical text. The Arabians later retitled it *Almagest*. The date of the first Arabic translation was probably 827.

With *Almagest* in their possession and now in their language, the Arabs had the basic foundations from which they could launch their own extensive inquiry into the science of the stars. But Ptolemy was not to be accepted blindly; and the great mathematician and astronomer, Al-Kwarizmi (780-850) was commissioned to verify and possibly update anything he found in Ptolemy's work. Al-Kwarizmi also drew up the first star tables in Arabic using the Indian system of counting—a system far superior to what the Arabs had been using. Arabian mathematics, which now incorporated this new counting system, became so successful that the Indian numeric system has since come to be known as the Arabic numbering system. Later, Al-Battani (850-929), also known as Abattegnius, would modify some of Ptolemy's calculations such as the true length of the year and the constant of precession. His work was translated into many of the European languages and became very popular among academics and scientists. Al-Battani believed that astronomy was second in importance only to religion.

In the early part of the eleventh century, a new version of an ancient altitude measuring device called the "astrolabe" was introduced. The astrolabe possibly dates back to Hipparchus or Appollonius of Perga. In its most simple version, the instrument was a wood or metal dish with the circumference marked to indicate degrees. The dish was usually hung from a ring. At its center was an adjustable pointer called an alidade. By using the alidade for sighting and the marked dish for positional readings, angular distances could be determined. The astrolabe was a forerunner of the sextant, invented in the eigtheenth century.

The eleventh century was also a time of a major compilation of new planetary and trigonometrical tables which eventually came to be called Toletan Tables. And in the century that followed, there were new translations of *Almagest*, publications of updated astronomical tables, and a brand new interest in Aristotle's cosmology. This last came as a result of the philosophical ideas of Avicenna (980-1037), the Arabian physician, and Averroes (1126-1198), the Spanish-Arabian philosopher. Both men were seeking to establish strong philosophical foundations for religious doctrines as well as a metaphysical argument for the existence of God. Both men were also fascinated with Aristotle's cosmology both from a physical and philosophical perspective; and as they each favored a cosmological (or causal) argument for the existence of God, as did Aristotle, they stirred new interest in the science and philosophy of Aristotle. Aristotle had argued that the causal sequence in nature was proof of a prime mover, an idea which Thomas Aquinas later included as one of his four arguments for the existence of God. As it would have seemed a contradiction in those times to praise Aristotle's philosophy and not his science, Aristotle's universe of transparent spheres was seen as a cosmology fully in tune with the Christian faith and the authority of the Catholic Church.

Meanwhile, numerous astronomers enhanced the Ptolemic tables and helped advance the science of astronomy, though, for the most part, any advances in astronomy came few and far between; but, then, the Arabs, to whom the torch had been passed, were noted for their meticulousness and their concern for reputation. By the latter part of the twelfth century, new magnitudes were determined for the stars, new computing methods were devised for detecting eclipses and conjunctions, and new observations were made of equinoxes and solstices. Still, there was little success in coming up with a better scheme of planetary motion than Ptolemy had devised, though the Arab astronomers remained generally dissatisfied with the idea of epicycles and deferences and, in the thirteenth century, Al-Tusi (1201-1274) made a well-respected though flawed effort to devise an alternative to Ptolemic cosmology.

By the thirteenth century, the Spanish dominated the science, making the great contributions to astronomy. The Muslim culture that had planted new roots in Spain back in the early part of the eighth century was now beginning to wane as a result of internal strife and foreign invasion. The result was a decline in Spanish astronomy, and a wrestling of astronomy from the Muslim cultures by other European countries. In the thirteenth century, there were some noted efforts. The Spanish king of Castile and Leon, Alfonso X (1221-1284), succeeded his father, Ferdinand III. While Alfonso was, for the most part, a political failure, historians credit him with having given great cultural stimulation to the art and science within his territories. Among the important compilations that he was responsible for commissioning was something which came to be called the Alfonsine Tables. These were mostly minor revisions to the Ptolemic star tables that were compiled and verified by a group of about 50 astronomers who were assembled in Toledo solely for the purpose of coming up with dependable planetary data. However, the tables, though completed in the year 1252, were not published until 1483. Meanwhile, in that span of time, the data the tables contained was disseminated in other ways to the scientific community, so when the tables were published, they were a bit of a disappointment, for the astrological and astronomical data the tables contained were already in the public domain. Still, the Alfonsine tables remained a primary source of reference for astronomers until the late seventeenth century.

## THE EARLY MODERN PERIOD

What is referred to here as "the early modern period" is really the golden age of discovery in astronomy. Four men launched it: Nicholas Copernicus, Tycho Brahe, Galileo Galilei and Johann Kepler.

Copernicus (Fig. 1-10) was a painstaking observer and theoretical genius. He was able to determine the relative distance of the planets from the Sun with almost perfect accuracy. But his main claim to fame was his daring—his daring to break with the past, to challenge the superstitions of the times. Copernicus (1473-1543) was a Polish astronomer, born in Torun and educated in astronomy at the University of Cracow (although he left that university in 1494 without taking his degree). However, his formal education did not end with his departure from Cracow. He went on to study canon law, medicine, philosophy, and law in Italy. By the time he completed his education and assumed duties as a resident canon in the Cathedral of Fraunberg in the Baltic (in 1512), he was mathematician, astronomer, jurist, and physician. His contribution in all of these fields was extensive but it was his work in astronomy that has immortalized him, especially his great literary effort *Revolutionibus Orbium Caelestrium* (Revolution of the Heavenly Spheres). In this masterpiece he developed a planetary scheme that had the Sun at its center and Earth, as one of the planets, revolving about it. Though he completed the text in 1530, he did not publish it until 1543.

Why did he wait so long? Of what was he afraid? He was particularly afraid of the Catholic Church and the personal scandal he would receive for challenging its authority. How could an astronomical treatise challenge the authority of the Church? Well, in Copernicus' time, the papacy called for literal interpretation of most passages in the Bible, and any individual who challenged that authority was publicly exposed as a heretic and a blasphemer. Copernicus knew he would have the support of the Church's intellectuals, those theologians who never saw the search for knowledge as a challenge to true faith. But the governing body of the Church, and those who were overly-sensitive about the political and physical survival of the Church and its members in a time of growing rivalry between Catholics and Protestants, were easily threatened. And the planetary scheme Copernicus had devised put the Sun at the center of the universe while current Biblical interpretation put Earth at its center. Copernicus was so afraid of the repercussions if his intentions were misunderstood that if it were not for the urging of close friends (some who were in responsible positions within the Church, such as Nicholas Schonberg (Cardinal of Capua) and Tideman Giese (Bishop of Culm)) he probably would not have dared to publish his ideas. As it turned out, the day of May 24, 1543, when Copernicus finally saw an advance copy of his treatise, was also the day of his death. Copernicus had been bedridden for sometime with apoplexy and paralysis.

In *Revolutions*, Copernicus stated that the universe is shaped very much like a globe and so, too, is Earth. That Earth is globe shaped, he explained, is evident because as people travel to the north from any direction, the northern vertex of the axis of daily revolution moves overhead to the same extent that the other moves downward. Northern stars are not seen to set and southern stars are not seen to rise.

Table 1-8. Early Modern History of Astronomy.

| Name | Comment |
|---|---|
| Nicholas Copernicus (1473-1543) | The Sun is the center of the universe and the Earth revolved around it; the universe is shaped like a globe and so, too, is Earth; there are three movements to Earth: the circuit of day and night, the annual movement, and the declination. |
| Tycho Brahe (1546-1601) | Made extremely accurate observations of the stars and planets; drew up star catalogs; disproved Aristotle's theory of transparent spheres; objected to Ptolemic cosmology but could not verify Copernicus' model of the universe. |
| Johannes Kepler (1571-1630) | Each planetary orbit is elliptical and the Sun is one of the foci; the line joining the planet's center with the Sun's center moves over equal areas in equal periods of time for all orbital points; wrote a book on comets; believed the universe is a harmonious system; put to rest once and for all Ptolemic astronomy and cosmology. |
| Galileo Galilei (1564-1642) | Constructed the first astronomical telescope, discovered four satellites of Jupiter, investigated sunspot phenomena and made countless contributions to astronomy and to telescope design. |
| James Gregory (1638-1675) | Designed and improved the reflective telescope to help alleviate some of the chromatic and spherical aberrations which were inherent in the original Galilean design. |
| Christiaan Huygens (1629-1695) | Improved the art and craft of telescope making; discoverd the rings of Saturn and one of its satellites. |
| Edmund Halley (1656-1742) | Used the transit of Venus to determine the Sun's parallax and was the first to predict the return of a comet. |
| James Bradley (1693-1762) | Discovered the aberration of light and the mutation of Earth's axis. |
| Johann Palitzch (1723-1788) | Discovered the comet later named for Edmund Halley. |
| Charles Messier (1730-1817) | Compiled a list of more than 100 nebulae; distinguished dust clouds from comets. |
| Joseph Lalande (1732-1807) | Director of Paris Observatory; popularized astronomy through his many writings. |
| Sir William Herschel (1738-1822) | Pioneered the study of the stars; designed and built his own telescopes; discovered Uranus; discovered the infrared range of the electromagnetic spectrum. |
| Johann Bade (1747-1826) | Developed a formula for determining the relative distances of the planets from the Sun; founded the Astronomical Yearbook. |
| John Delambre (1749-1822) | Developed a table on the motions of Uranus. |
| Caroline Lucretius Herschel (1750-1848) | Verified many of Sir William Herschel's observations; discovered nebulae and comets and produced and published star catalogs. |
| Stephen Groombridge (1755-1832) | Established that comets were extraterrestrial objects and produced extensive star catalogs (along with Ernst Chaldni). |
| Heinrich Olbers (1758-1840) | Discovered the asteroid Palla and many comets; believed asteroids are products of a collision between a planet and some other massive object. |
| William Wallaston (1766-1823) | Discovered the dark lines in the solar spectrum, therby establishing that certain colors in the solar spectrum do not blend together. |
| John Herschel (1792-1871) | Discovered many double stars and verified many of Sir William Herschel's discoveries. |
| Friedrich Bessel (1784-1846) | Discovered the parallax of 61 Cygni and increased the number of known stars to over 50,000. |
| William Cranch Bond (1784-1859) | Used photography for stellar observations; became the first director of the Harvard Observatory; contributed greatly to knowledge about the Sun, nebulae, and planets. |
| John Encke (1791-1865) | Successfully determined Mercury's mass and determined the motions of many stars. |
| Friedrich Struve (1793-1864) | Predicted that there must be light-absorbing systems between the stars, was one of the first to obtain a stellar parallax. |
| Sir George Biddell Airy (1801-1892) | Contributed to the development of improved astronomical instruments through his research on magnetism and meteorology. |

| | |
|---|---|
| Pietro Secchi (1818-1878) | Aided in the development of spectroscopy; pioneered the technique of identifying stars by their spectra. |
| Sir William Huygens (1824-1910) | Determined the chemical composition of stars and pioneered the use of spectroscopic telescopes. |
| Simon Newcomb (1835-1909) | Contributed greatly to knowledge about the orbits of Mercury, Venus, Mars, Earth, Uranus, and Neptune and developed formulas for explaining lunar motions. |
| Giovanna Schiaparelli (1835-1910) | Argued that meteor swarms follow cometary orbits; showed that Mercury and Venus both rotate on their axis; and showed there were canal-like structures on the surface of Mars. |
| Agnes Mary Clark (1842-1907) | Produced important works on astronomy and astrophysics. |
| George Ellery Hale (1868-1938) | Organized the Mt. Palomar, Yerkes, and Mt. Wilson Observatories and invented the spectroheliograph. |

Additionally, polar inclinations from every point have the same ratio with places equally distant from the poles—and that cannot happen in any but a spherical figure. Further proof comes from those who sail the seas. When land is not visible from the deck of a ship, it is still visible from the top of the mast; conversely, if something shines from the top of the mast, it appears to those onshore to come down gradually as the ship moves further out to sea until finally it appears to set behind the horizon, as the Sun would set.

Fig. 1-10. Nicholas Copernicus (1473-1543) was the Polish astronomer, lawyer and doctor who laid the basis for a heliocentric theory of planetary motion (Courtesy of Yerkes Observatory).

There are three movements of Earth, Copernicus explained in the 11th chapter of Book 1. The first is the circuit of day and night which goes around the axis of Earth from west to east and describes the equator or what is called the "equinoctial circle." The second is the annual movement of the center which describes the circular paths of the zodiacal signs around the Sun from west to east. The third is the declination, also an annual movement but a movement toward the signs which precede (from Aries to Pisces), or westwards; i.e., turning back counter to the center movement; and, in consequence of these two movements which are almost equal to each other but in opposite directions, it follows that the axis of Earth and the greatest of the parallel circles on it, the equator, always look towards approximately the same quarter of the world, as though they remained immobile." Earth, in turn, he explained, rotated as all other wandering stars (the planets) about the Sun (Fig. 1-11). And these ideas were his major challenge against the religious traditions of the sixteenth century Catholic Church: the rotation and revolution of Earth, and the Sun at the center of the known universe.

The unfortunate aspect of Copernicus' work was that it continued to support Ptolemy's use of epicycles and deferents to explain planetary movement; at least Copernicus broke free of the Ptolemic idea of a geocentric model.

He did something else, however, which was a third challenge to Catholic ideas of the time. He explained that Earth was an insignificant world in the grand scheme of things. The theological implications

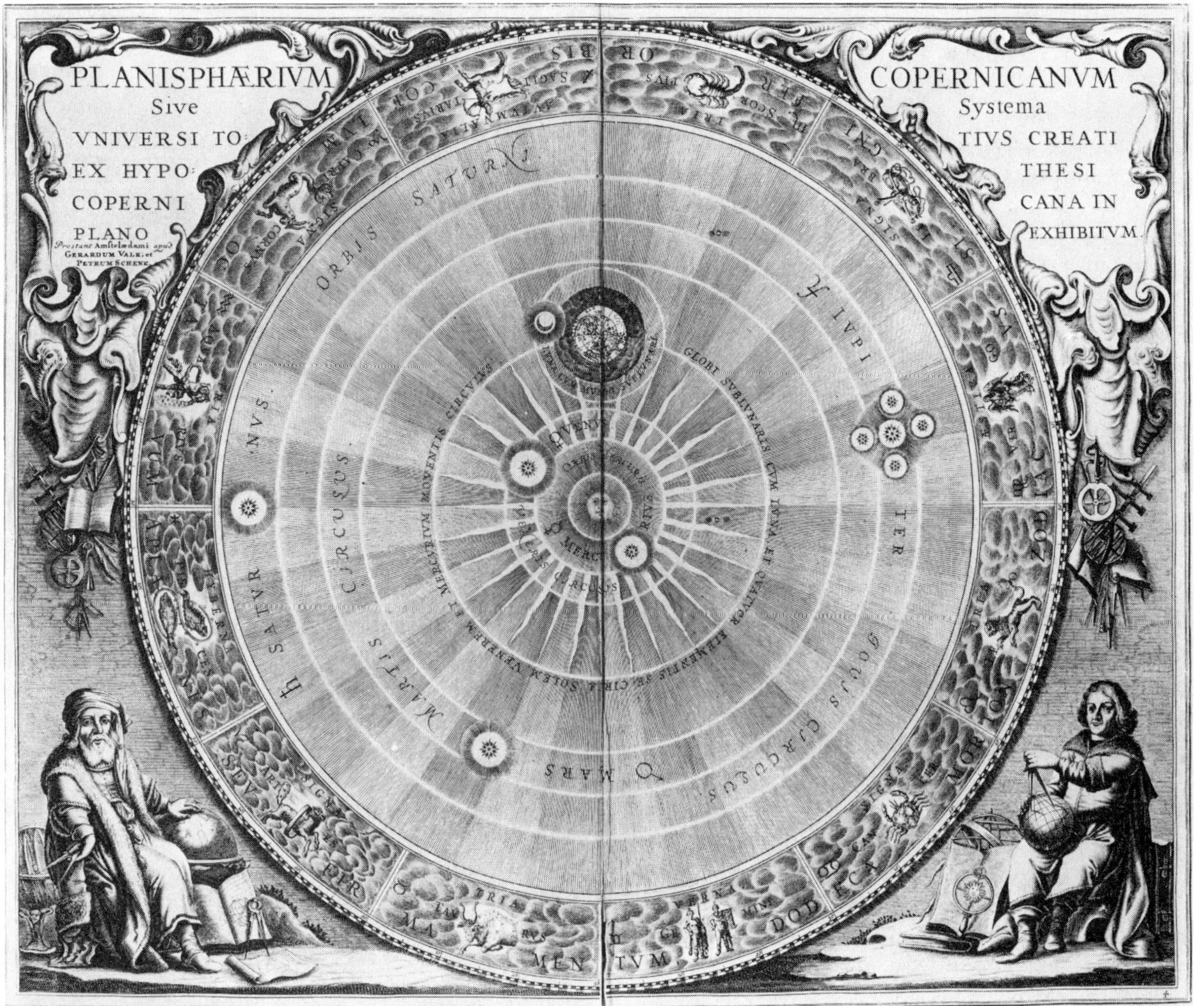


Fig. 1-11. Copernicus put the Sun as a motionless body at the center of the universe and had Earth and the other known planets orbiting about it.

here were astounding and he knew it, for if man was God's greatest creation, He would not have given man an insignificant world on which to live.

As it turned out, all of Copernicus' fears were unnecessary. His ideas did receive some criticism but this was offset by support from many astronomers and physicists, as well as from his own colleagues and friends. As for the possibility of excommunication or social and professional embarrassment, the odds were against it. Generally speaking, the Church showed no hostility toward his ideas. This was for two reasons. The first was the strictly technical approach he used in describing his world system; it helped portray him as a serious scientist in search of truth rather than as a critic of Scripture. His approach was very mathematical and very logical, and those with the education and interest to study his ideas were left only with a great deal of admiration for his logical arguments. The second was a preface, probably not written by Copernicus but possibly by a close associate or supporter, in which it is explained in so many words that the ideas presented in the book are given in full respect for religious beliefs and those who might object to these ideas on the basis of merely traditional grounds, "by shamelessly distorting the sense of some passage of Holy Writ," must see their judgments as merely foolishness and their attitudes subject to scorn. Later, Galileo would echo Copernicus' ideas, but the times would be different and an easily-threatened Church would brings its wrath down upon him. Before discussing Galileo's sad fate, there are three other great men in this chronology who need mentioning: Giordano Bruno, Tyco Brahe, and Johannes Kepler.

Bruno (1548-1600) was an Italian philosopher and former Dominican who lacked Copernicus' political sense. He is a perfect example of what Copernicus may have feared. Bruno was not an astronomer but as a philosopher would necessarily have a deep interest in cosmogony and cosmology, disciplines concerned with the origin and order of the universe. Bruno stood fast against dogmatism of any kind; he felt that the ideas of men were too relative to their immediate view in space and time, and as both of these last changed constantly, the search for real truth must be a continuing enterprise. Dogmatic teachings were by their very nature, therefore, destined to become outmoded and useless; the human mind must never be tied in a Gordian Knot. Bruno believed that the world was composed of very basic elements called monads; these cannot be seen or detected in any way but they interrelate in such a way that they cause the order of the universe. In fact, according to Bruno, the very existence of the universe is owed to these monads, which are governed by some supreme force. Bruno became involved in astronomy mainly because he was intrigued with the Copernican world system and though he objected to many of Copernicus' other ideas, he actively supported the fundamental ideas in *Revolutions*: that Earth was not the center of the universe, that it was just a minor member of a massive stellar population, and that man was an insignificant speck in the universe. These ideas added to many of his outspoken attacks on the Church and its traditions, led him to be tried for heresy during the Inquisition. This was not the first time such a charge was brought against him, for as a young Dominican in 1576 he was forced to flee his order because of his outspoken challenges to Catholicism. This second time there was no escape; his trial resulted in his being imprisoned and then burned at the stake, as an example to all those who would challenge Church dogma with pagan science and ideas. Galileo was ever mindful of the fate that befell Bruno.

Tyco Brahe (1546-1601) was born about two years after Bruno had escaped from those first charges of heresy by the Dominicans and had entered his second career as itinerant teacher. Brahe (Fig. 1-12), however, was quite a different man than Bruno. He had little interest in philosophy; he was an astronomer, first, last and always; and he was to become the most noted astronomer of his time. A Dane, he built his own observatory at Stjarneborg in 1584. Rather early in his career he became noted for his exceptional technical abilities in improving as-

Fig. 1-12. Tyco Brahe (1546-1601), the Danish astronomer, made uncannily accurate observations of the planets. His records served as the basis for Kepler's laws. Kepler admired him, but he never trusted Kepler, and was always afraid Kepler would steal the credit for the work he had done. On his deathbed, however, he begged Kepler not to let the world forget him (Courtesy of Yerkes Observatory).

tronomical instruments as well as for his keen observational talents. His extremely accurate observations of the planets, observations which Johannes Kepler would later use as the basis for his laws on planetary motion, were uncanny in their precision. Brahe and his staff drew up new star catalogs, once and for all shattered Aristotle's idea of transparent spheres, and discovered new facts about the orbital characteristics of the Moon. Tycho could never verify the Copernican model of the universe; neither could he accept the Ptolemic universe. So he struck a compromise in which the five known planets (Earth was not accepted as a planet) circled the Sun, and the Sun, in turn, circled Earth.

Brahe's work came to almost a dead-halt in 1588 when his benefactor, King Frederick, died. Because of his many quarrels with many individuals in strong political and social standing, he was not able to argue successfully for either a continued allowance or a pension. The years that followed Frederick's death found Brahe struggling to survive and he was forced to continue his work from a number of addresses in Copenhagen, Rostock, Wandsbeck, Wittenberg, and finally Prague. It was in Prague that he met Johannes Kepler.

Kepler (1571-1630) was a German astronomer who had been greatly influenced by the ideas of Copernicus. In 1596, he authored a work called *Mysterium Cosmographium*, which opened the way for correspondence with Brahe and Galileo. One result of this correspondence was Kepler's joining Brahe as his assistant. At a castle just a few hours from Prague by carriage, Kepler labored diligently beside Brahe. Brahe could never get himself to fully trust this young genius who showed incredible mathematical ability. Brahe had never been a man to cultivate friendships or to show respect for others; people reacted to him, then, as might be expected: they reacted coldly to him or avoided his company unless it was absolutely necessary for them to tolerate his presence. Since the death of Frederick, Tycho had no loyal patron and had had to suffer eleven years of relocations and re-establishment. By the time Kepler joined him, Tycho was already afraid the younger man would steal his notes and take credit for most of the startling observations that Tycho had made and recorded in his journals. Yet, when Tycho, at age fifty-five, lay on his deathbed, he begged Kepler not to let the world forget him. He hoped that Kepler would complete the work they were doing on the motions of Mercury, Venus, and Mars in full accordance with the system he, Brahe, had established.

Brahe should never have feared Kepler's intentions. Kepler was a dedicated scientist and man of the highest moral character. It was highly unlikely that he would ever fail to give credit where credit was due. In 1609, he published the results of Brahe's research on the orbital patterns of Mars. In this work, he stated the first two of his famous laws of planetary motion: that each planetary orbit is elliptical and that the Sun is one of the foci, and that the line joining the planet's center with the Sun's center (the radius vector) moves over equal areas of the elliptical orbit in equal times. In 1618, he wrote a book on comets. A year later, greatly influenced by Pythagorean ideas, he wrote and published a book on the harmonies of the world; in it, he argued that the universe is a harmonious system governed by strict mathematical laws. The idea that the universe is a harmonious system is now considered absurd, but in the book Kepler developed the third of his famous laws concerning planetary motion: the square of the time in which a planet completes its orbit is in proportion to the cube of its mean distance from the Sun. What was most important about Kepler's work was that it put to rest astronomy's apologetic attitude toward Ptolemic astronomy.

While Kepler was still hard at work putting Brahe's observations into formulas, the Italian astronomer, mathematician and physicist, Galileo Galilei (1564-1642), had heard that a Dutch lens maker by the name of Hans Lippershey had been experimenting with ways in which very distant objects could be magnified. Lippershey was a businessman and not a scientist and his main interest was in turning his invention into a highly profitable investment; he had little foresight or even interest in what his invention might mean to astronomy. Galileo, on the other hand, saw the great advantage Lippershey's invention would have to astronomers. By 1609 Galileo had constructed the first astronomical

Fig. 1-13. Galileo demonstrates his telescope to a fascinated public. Galileo (1564-1642) did not actually invent the telescope but he was the first to use it for astronomical investigations. He made many startling discoveries, including what are now known as the Galilean satellites of Jupiter (Courtesy of Yerkes Observatory).

telescope (Fig. 1-13). Within the year that followed he was able to construct a telescope with a magnification power of 33 diameters. His telescope was not much more than a long tube with a concave eye glass and convex object glass but once he pointed it to the skies, it became a priceless tool. Galileo discovered that the planets were much more than anyone had ever imagined and that the Moon was indeed a world unto itself. His telescope would herald a new age of discovery in astronomy. Men might never be able to travel to the planets, but with the telescope they could bring the planets to them.

In January of the year following the construction of his first telescope, Galileo discovered four satellites of the planet Jupiter, studied the phases of the planet Venus, and investigated sunspot phenomena. Most importantly, he was able to confirm Copernicus' theories, though he kept this finding to himself, for the Copernical system was now considered in opposition to Church teaching. Gradually, Galileo began to reveal his findings (though never formally) mainly through his teaching and other academic communications while at the University of Pisa. In 1616, the Church officially opposed the Copernical world system and Galileo, who had demonstrated his telescope to a papal audience in 1611, was now summoned to Rome and warned not to give any further support to Copernican ideas.

For about 16 years, Galileo wrestled with his responsibilities as a scientist and with his responsibilities to the Church. Finally, he could no longer hold back the truth and in 1632 he published a text which upheld the findings of Copernicus. Guardians of Church doctrine were enraged and Galileo was once again summoned to Rome, this time to be tried by the Inquisition. Galileo had been wise and respectful enough to have sent his manuscript to Church censors before proceeding with publication, and while they had recommended some changes in the way he presented some of his ideas, they were unaware that the text so heartily supported Copernicus. When the work was published and received with great applause, the Church interpreted the book and its supporters as openly ridiculing Church teachings. The censors who had edited the book were also on the spot. They could very well have been seen to be part of the "conspiracy," for they were well aware, as Galileo was, that in 1616 there was an ecclesiastical decree condemning the Copernical world system.

The Inquisition found Galileo guilty of claiming the Holy Scriptures to be false, for he believed the Sun to be the center of the universe and Earth in motion. His punishment was imprisonment at his home in Arcetri—a rather lenient punishment in a time when people could be burned at the stake for heresy, or have their eyes gouged out. It has always been an amazement to theologians, philosophers, and historians that a religious institution which tied its existence to a man-God who never believed the end justified the means, who never sought to harm even his enemies, would engage throughout much of its history in such severe campaigns to win and maintain allegiance by methods hardly in accordance with what would seem to be the moral order. But in Galileo's time the Church was just as much a political entity as it was a religious institution, and it was struggling for survival during a time of political turmoil and numerous and growing protestant sects. It lost its vision and its purpose amidst the challenges.

Galileo became very ill after his trial and eventually went blind. He must have been a very disillusioned man in his last years, yet he kept up his work even after his blindness. He must have felt some comfort in knowing that others would confirm his findings; he knew that the truth could not be hidden forever. With his telescope, men and women would be able to gather observational evidence that could not be refuted simply because of traditions, interpretations of scriptures, or religious authority.

The telescope that Galileo had constructed was the refracting type. A lens on the far end of the tube served as the means for collecting available light. A lens at the near end of the tube served as the means of magnification. It was a fantastic instrument for its time but its design resulted in chromatic and spherical aberrations, problems which have since been overcome by newer design concepts and lens combinations. This Galilean version was the only type of astronomical telescope until James Gregory (1638-1675), a Scottish mathematician, designed a reflecting telescope in which the collecting mirror of the refracting telescope is replaced by a mirror, and viewing can be done from the side of the telescope. Isaac Newton (1642-1727), whose work on the physics of light is discussed in Chapter 2, seized Gregory's idea and developed it even further.

By the seventeenth century the rapid progress being made in all sciences began to influence the way in which the papacy began to look at the role of science. Many Church theologians and philosophers were already willing to leave science to take its own course but this was not the reason the Church began to change its stand. It was, rather, that the Church's congregations were increasingly more educated and the people themselves were thoroughly interested in the progress of science. Additionally, there was a new theology and philosophy taking root; it had its foundations in teleology—or in evidences of design in nature, which is basically what teleology is all about. Natural theologies and natural philosophies sprung up from all quarters; together they argued that all the cosmos, as well as the humble planet Earth, was significant indication of an orderly scheme, of design, and wherever there is design there must have been a designer. God was now looked upon as a great scientist who designed the entire universe according to strict physical laws which men and women could sense, investigate, and learn. By studying the natural order of things, men and women could find the practical morality which must guide them as well as a practical religion to support that morality.

The idea that astronomy, specifically, would add to one's knowledge can actually be traced back to Old Testament writings. In the books of *Genesis, Job,* and *Jeremiah*, the idea that men and women can come to know God by understanding the heavens is clearly stated, as it is in the New Testament book *First Corinthians*. The same idea appears in Plato's *Republic* and in his *Timaeus*, in Aristotle's *Physics, Metaphysics,* and *Heavens,* in Lucretius' *Nature of Things,* Aurelius' *Meditations* and Tacitus' *Histories*.

In the seventeenth through the nineteenth centuries, astronomy remained in the domain of the philosopher as well as the scientist. Metaphysicians such as Immanuel Kant (1724-1804), perhaps the greatest name in philosophy since Thomas Aquinas, continually surveyed astronomical and cosmological findings to support their ideas, though, of course, in the case of the great thinkers like Kant, their philosophical arguments were not in any way dependent upon simply the weight of observational evidence. Kant himself readily recognized the logic underlying the Copernican cosmology and likened his transcendental philosophy to the Copernican astronomy in that they both represented great breaks with the past.

While the great philosophers of the eighteenth and nineteenth centuries looked to astronomy for teleological evidences of God and the possible immortality of men or to cosmology for the causal sequences that would at least identify the existence of a First Mover, amateur and professional scientist alike continued to study the skies and make astonishing discoveries. Many of those made in the late seventeenth century through the eighteenth century belonged to the Herschel family.

Sir William Herschel (1738-1822) pioneered the study of stars (Fig. 1-14). Originally a musician, he gradually developed such a preoccupation with astronomy that it became his full-time occupation. He designed and built his own telescopes and became so appreciated for his work and detailed observations, which included the discovery (in 1784) of the

Fig. 1-14. William Herschel (1738-1822) started out as a professional musician but eventually became so thoroughly obsessed with astronomy that the latter became his full-time occupation. He served as the king's private astronomer (Courtesy of Yerkes Observatory).

planet later to be called Uranus, that he was appointed the king's private astronomer. After he had developed his famous telescope (Fig. 1-15), designed for a 48-inch mirror and 40-foot focal length, he proceeded to catalog more than 800 double stars,[25] discovered satellites of Saturn and conducted a great deal of research on nebulae. He developed a model of the Milky Way and theorized that nebulae were the birthplace of stars. He also surmised that as the planets circled about the Sun, the Sun, in turn, led them forward through space.[26]

Sir William was fortunate to have as his collaborator on many projects beginning in 1772, his sister, Caroline Lucretius Herschel (1750-1848). She

Fig. 1-15. A sketch of Herschel's 40-foot telescope (Courtesy of Yerkes Observatory).

not only verified or commented on her brother's observations but also made her own important contributions and discoveries (nebulae and comets), and produced and published star catalogs (Fig. 1-16).

Sir William's son, John (1792-1871), noted also for his contributions to chemistry and photography, verified much of his father's findings about double-star populations and discovered many additional double-stars which he added to existing catalogs. Much of his early research was conducted with Sir James South (1785-1867), the English astronomer who was the founder (in 1829) and first president of the Astronomical Society.

But concentrating on the Herschels brings us too quickly through the eighteenth and nineteenth centuries and leaves us to miss many of the noted astronomers who lived in the 1600's and 1700's (Fig. 1-17). Among these famous men must be included the Dutch mathematician, Christiaan Huygens (1629-1695), who greatly improved the art and craft of telescope lens making, and who was the first to discover that Saturn was surrounded by rings (Fig. 1-18). He also discovered one of the satellites of Saturn and, like Newton, did important research on the physics of light.

Another seventeenth century astronomer whose name is remembered today is Edmund Hal-

Fig. 1-16. Caroline Lucretius Herschel (1750-1848) discovered many nebulae and comets and produced and published star catalogs (Courtesy of Yerkes Observatory).

ley (1656-1742), the first astronomer to both predict the return of a comet and to use the transit of Venus to determine the parallax of the Sun. He was also the man who subsidized the publication of Newton's *Principia* and lent his expertise to help prepare it for the press. The great comet of 1682, whose orbit he had predicted and which has come to be known as Halley's comet, was not actually his discovery. The true discoverer of that great wanderer of the solar system was a German astronomer and farmer by the name of Johann Palitzch (1723-1788) who spotted it on Christmas Day in 1758.

Like Huygens and Newton, there was another astronomer who was equally fascinated with the physics of light as well as the science of astronomy: James Bradly (1693-1762), the Oxford-educated astronomer who would become astronomer royal in 1742. Bradly discovered something called the "aberration of light" as well as something referred to as the "nutation of the Earth's axis." The first term refers to the apparent position of a star which will change according to the direction in which the Earth moves relative to it. The second term refers to a periodic but irregular motion on the Earth's processional circle caused by the Sun and the Moon.

During the time of the Herschels, there were a great many discoveries by other astronomers. There were the cometary discoveries of the French astronomer Charles Messier (1730-1817) as well as his compilation of over a hundred nebulae, a task he had begun mainly to distinguish the dust clouds from being mistaken for comets;[27] the star maps and exhaustive catalog of stars (over 17,000 stars) and nebulae published in 1801 by the German astronomer, Johann Bade (1747-1826), his famous formula for determining the relative distances of the planets from the Sun, and his founding in 1774 of the *Astronomical Yearbook*.

In addition, there were the important contributions of John Delambre, Stephen Groombridge, Ernst Chladni, and John Goodriche.

Delambre (1749-1822) was a French astronomer and mathematician who became famous not only for his many astronomical computations but also for developing a table of the motions of Uranus. Stephen Groombridge (1755-1832) distinguished himself with his extensive star catalogs and Ernst Chladni made his reputations through his studies of comets and by identifying them as extraterrestrial objects. John Goodriche (1764-1786) earned his fame for his studies of star luminosity and brightness variations.

In the first quarter of the nineteenth century, the names which came to the forefront of astronomy were William Wallaston, John Dalton, Guiseppe Piazzi, Heinrich Olbers, Joseph Lalande, Joseph von Fraunhofer, and Heinrich Schumacher.

William Wallaston (1766-1828) was an English scientist known for his invention of an instrument used for measuring the angles of crystals (reflecting goncometer), for his discovery of certain elements (palladium and rhodium) and for his many accomplishments in geology and other natural sciences. In astronomy, he made news by his dis-

I

# STELLARUM
## INERRANTIUM
## CATALOGUS BRITANNICUS,

Ad Annum Chrifti Completum, 1689.

Ab Obfervationibus GRENOVICI in OBSERVATORIO Regio habitis.

Affiduis Vigilijs, Cura, & Studio

*JOHANNIS FLAMSTEEDII*, Aftronom. Reg.

Deductus & Supputatus.

In Conftellatione ARIETIS.

| ORDO Ptol. | ORDO Tych. | STELLARUM Denominatio. | Bayer Cha. | Afcenfio Recta. ° ' " | Diftantia à Polò B. ° ' " | Longitudo. s ° ' " | Latitudo. ° ' " | Varia. Afc.R. ' " | Varia. D. à P. ' " | Magnitud. |
|---|---|---|---|---|---|---|---|---|---|---|
| | | | | 20 46 0 | 69 17 15 | ♈ 26 58 25 | 11 4 58 B | 58 7 | 22 25 | 7.6 |
| | | | | 21 25 45 | 71 15 25 | 26 48 15 | 9 1 26 B | 57 52 | 22 19 | 7.6 |
| | | | | 22 26 15 | 74 10 55 | 26 36 18 | 5 57 3 B | 57 31 | 22 12 | 6 |
| | | | | 22 51 15 | 74 36 55 | 26 49 4 | 5 23 59 B | 57 28 | 22 07 | 7.6 |
| 1 | 1 | Quæ in Cornu duarum præcedens | γ | 24 8 30 | 72 14 45 | 28 51 0 | 7 8 58 B | 58 02 | 21 55 | 4 |
| 2 | 2 | Sequens & *Borea* eft | β | 24 23 30 | 70 43 55 | 29 37 59 | 8 28 16 B | 58 22 | 21 50 | 3 |
| | | | | 24 39 45 | 67 57 45 | ♉ 0 54 20 | 10 57 12 B | 58 57 | 21 49 | tel. |
| 5 | 6 | In Cervice | ι | 25 7 0 | 73 43 15 | ♈ 29 10 57 | 5 26 12 B | 57 54 | 21 44 | 6 |
| | | *In Vertice* | λ | 25 11 0 | 67 56 [illegible] | ♉ 1 22 15 | 10 47 47 B | 59 00 | 21 44 | 5 |
| | | | | 26 32 [illegible] | 65 35 5 | 3 26 14 | 12 31 52 B | 59 45 | 21 28 | 6.7 |
| | | | | 27 19 30 | 65 48 15 | 4 2 12 | 12 4 2 B | 59 48 | 21 20 | 6 |
| | 17 | *Infra Lucidam* | κ | 27 19 30 | 68 51 5 | 2 55 8 | 9 13 29 B | 59 10 | 21 21 | 6.5 |
| *Inf.* 1 | 3 | Infor.fup.Caput, *Lucida* ♈*tis* | α | 27 26 30 | 68 1 45 | 3 19 18 | 9 57 12 B | 59 22 | 21 20 | 2 |
| | | | | 27 57 30 | 65 33 15 | 4 40 46 | 12 5 32 B | 59 56 | 21 12 | 6 |
| | | | | 28 22 30 | 71 58 45 | 2 43 49 | 5 56 58 B | 58 36 | 21 08 | 6 |
| | | | | 28 24 45 | 65 32 40 | 5 4 35 | 11 57 0 B | 60 4 | 21 06 | 8 |
| 3 | 4 | In Roftro duarum Borea | η | 28 52 30 | 70 16 25 | 3 46 50 | 7 22 45 B | 59 2 | 21 02 | 6 |
| | | | | 28 58 45 | 71 33 15 | 3 25 14 | 6 8 45 B | 58 46 | 21 00 | 7 |

A

Fig. 1-17. A page from a star catalog written by John Flamsteed (1646-1719), a contemporary of Halley, Huygens, and Newton. The page is from his "Catalogus Britannicus" in the *Historia coelestis Britannicus* published in 1689 (Courtesy of Yerkes Observatory).

Fig. 1-18. Christiaan Huygens (1629-1695) was a Dutch mathematician who contributed greatly to the art and craft of telescope making. Among his discoveries was one of the satellites of Saturn (Courtesy of Library of Congress).

covery of the dark lines in the solar spectrum, a rather important discovery in his time, for even the best experiments of Isaac Newton and Johann Wilhelm Ritter (1776-1810) had not revealed that certain colors in the solar spectrum do not actually blend together but are, rather, separated with dark lines. These lines were later named after another astronomer, Joseph von Fraunhofer (1787-1826), a German optician and physicist who was the first to thoroughly investigate and map these strange spectral lines.

Meanwhile, as Wallaston was pursuing the many fields that tempted his genius, John Dalton (1787-1826), who had in the last decade of the eighteenth century published essays on meteorology and was now studying the reasons for his own affliction (color blindness), was also doing extensive theoretical work on atomic particles. By 1808, he published the first of his two volumes on *A New System of Chemical Philosophy*. For centuries, the scientific community had been in hot debate over the existence of some chemical property that formed the basis of any existent, a chemical property that was unable to be divided or broken down in any way. In his book, Dalton had developed his law of multiple proportions and in doing so added new force to those arguments for the existence of such a basic substance as the atom was supposed to be. Today, we learn that physicists have split the atom; this is a contradiction. Such a possibility only means that science has not yet discovered the atom or else are calling something else by that name.

In terms of astronomical discoveries or theories, the nineteenth century got off to an important start (depending upon whether one considers the year 1800 belonging to the eighteenth or nineteenth century). First, there was William Herschel's discovery of infrared rays in the solar spectrum and then Guiseppe Piazzi's discovery of Ceres, the first asteroid to be discovered. Piazzi was a member of a Catholic (Theatine) order of priests which had been founded in 1597 for the purpose of combating the swelling tide of Lutheranism and which had hoped to set the foundations for a new moral order based on fundamental Christian teachings. Piazzi joined the order in 1765 and later became a professor of mathematics at the University of Palermo. But the discovery of Ceres was not his only contribution to astronomy. Piazzi also published a catalog of stars that in its final version included more than 7500 entries.

A second asteroid, Pallas, was discovered in 1802, the year after Piazzi's discovery, by the German physician and astronomer, Heinrich Olbers. In 1804, the asteroid Juno was discovered and three years later a fourth, Vesta; this last again by Olbers. Though he was first and foremost a physician, astronomy was much more than simply a passing interest. In 1779, he developed a system for calculating cometary orbits and proceeded to discover a number of comets. One has been named after him. Additionally, he developed some heavyweight theories about asteroids which are still under consideration today. He believed that at one time there existed another planet in our solar system but this planet collided with another body—a moon, comet or possibly another planet—or just exploded. The asteroids we find in the sky today are fragments of this lost planet, and they revolve around the Sun as the planet once did.

In 1807, the year that Olbers discovered Vesta, the noted French astronomer, Joseph Lalande (b. 1732) died. He was a professor at the College de France from the year 1761-1807, becoming in 1768 the director of the Paris Observatory. Lalande greatly popularized astronomy with his many books and also added to astronomical data through his technical writings and catalogs. His famous bibliography of astronomy had been made available to the public just four years before his death.

Another teacher and professor who contributed greatly to the literature on astronomy during this same period was Heinrich Schumacher, the Danish astronomer who was a director of the observatory in Copenhagen. Among Schumacher's many published efforts was a periodical on astronomical research and theory; it was launched in 1821 and called *Astronomische Nachricten*.

In the next quarter century (to 1850), new theories, discoveries, and technology were coming from both sides of the Atlantic with great frequency.

One American contributor was by trade a silversmith and clockmaker like his father. Born in Maine but settled in Boston, his name was William Cranch Bond (1784-1859). Astronomy was his main passion. He converted much of his home into an observatory, to which he would retreat at every opportunity. His reputation as both a technician and a scientist were so respected that he was commissioned by Harvard College to visit a number of observatories in Europe and gather whatever information was needed to build the best at Harvard. When the observatory at Harvard was finally completed, Bond became its first director, and using its instruments and library he proceeded to add a wealth of data to what was known about the Sun, nebulae, the planets, and the use of photography for stellar observations.

In Europe, Friedrich Bessel (1784-1846), a German astronomer, who was director of the observatory at Konigsburg since 1810, was attempting to find a way to measure the distance of stars. In 1838 came the announcement that he had discovered the parallax of 61 Cygni. This discovery represented a first; it was the first distance measurement of a star ever to be authenticated. Bessel continued to examine the skies until his death and succeeded in increasing the number of known stars to over 50,000. His work was continued by Friedrich Argelander (1799-1875), who published (in 1862) a catalog listing the position and brightness of more than 300,000 stars!

In 1842, Christian Doppler, an Austrian physicist formulated what has become known as the *Doppler principle*. The principle states that the frequency with which waves reach an observer increases or decreases depending upon the speed at which the distance between the two is increasing or decreasing. The actual effect which is observed on this frequency as the result of increasing or decreasing distance is called the *Doppler effect*. In simple terms, this means that there is an increase in pitch when the distance between two points decreases. In a study of the stars, which is a study of visible and invisible star light, how does this help in determining the relative motion of stars? Well, light waves and sound waves are similar. If the Doppler principle is applied to light waves, it is apparent that as the distance between a star and its observer decreases, there is a shift in one direction of the spectrum, and when distance increases there is a shift in the other direction.

To highlight the developments in astronomy from 1851 to 1875, let's discuss the accomplishments of the following fifteen men.

Johann Encke (1791-1865), a German astronomer, was director of the Berlin Observatory (Fig. 1-19). He successfully determined Mercury's mass by using as a point of reference the perturbations of a comet discovered by the French astronomer J.L. Pons (who discovered some 37 comets in a 26-year period beginning in 1801) but named after Encke. This was an exceptionally short-period comet (3.3 years), probably the shortest on record. Encke was honored instead of Pons in the naming of the comet because it was Encke who calculated its orbit and predicted its return. His calculations eventually led to the formulas which

Fig. 1-19. Johann Encke (1791-1865) was a director of the Berlin Observatory who distinguished himself in a number of areas in astronomy. The comet Encke was named after him, although the comet's discovery was actually made by J.L. Pons (Courtesy of Yerkes Observatory).

established the Jovian system of comets. Encke also calculated the distance from the Earth to the Sun based on the transit of Venus. His final calculations for Mercury's mass and Earth's distance were not accurate by any means, but they were surprisingly close. Encke also distinguished himself by determining the motions of a number of stars.

Friedrich Struve (1793-1864), was the director of the Dorpat Observatory from 1817-1829, and later director of the Pulkovo Observatory. Struve predicted that there must be light-absorbing systems between the stars, a prediction that has since been verified. This light-absorbing material had, with the advent of astro-photography, revealed the Milky Way to have an unusually varied distribution of stars. This revelation was simply the result of an illusion caused by scattered clouds, each cloud millions of miles or more in diameter, which float between stars and absorb much of the light from stellar regions. Struve was also one of the first astronomers to independently obtain a stellar parallax; and in 1838 he successfully measured the parallax of Vega. His son and grandson were also astronomers.

Friedrich Argelander (1799-1875), was a German astronomer and director (1837) of the observatory at the University of Bonn. He continued to add to Bessel's catalog of star positions. The results of his observations, along with those of Eduard Schonfeld (1828-1891) appear in a work called *Bonner Durchmusterung*, which describes the positions and brightness of more than 300 thousand northern stars.

Sir George Biddell Airy (1801-1892) was an English astronomer, whose research on magnetism and meteorology led to the development of new or improved instruments of measurement, among which were instruments that corrected compass errors in iron-built vessels and which improved telescopic observations ordinarily disturbed by atmospheric chromatic dispersion.

Urbain Jean Joseph Leverrier (1811-1877), Johann Gottfried Galle (1812-1910), and John Couch Adams (1819-1892) were preoccupied with the perturbations of Uranus. Leverrier's calculations indicated the probable presence of a planet somewhere beyond the orbit of Uranus. Meanwhile, separate observations by John Couch Adams resulted in the same conclusion. Galle, already the discoverer of three comets, used Leverrier's calculations to actually find the unknown planet—Neptune.

Christian Henry Frederick Peters (1813-1890) was an American astronomer. His discoveries included two comets and dozens of asteroids. He published a series of celestial charts over a six year period beginning in 1882.

Johann von Lamont (1805-1879) was the Scottish-German astronomer who made important discoveries in terrestrial magnetism. Besides making magnetic surveys in Europe from 1849-1858, he also proposed, in 1850, the magnetic decennial period. He is also famous for his studies of Uranus, for his cataloging of almost 35,000 stars, and for his measurements of nebulae and clusters.

Other major contributors were Pietro Angelo Secchi (1818-1878), the Jesuit priest who helped in the development of spectroscopy and pioneered the technique of classifying stars by their spectra; Hermann Ludwig von Helmoltz (1821-1894), the German scientist, physician, and philosopher who gave mathematical formulation to the law of conservation of energy; Peter Andreas Hansen (1795-1874), the Danish astronomer who revised lunar theory and compiled a book of lunar tables which were of great assist to navigators; Sir William Huggins (1824-1910), the English astronomer who determined the chemical composition of stars and pioneered spectroscopic photography; George Phillip Bond (1825-1865), the son of William Cranch Bond (the first director of the Harvard Observatory) and a pioneer in the use of photography for mapping the sky and determining the relative brightness of planets, and successor to his father at the observatory; John William Draper (1811-1882) and his son, Henry Draper (1837-1882), American scientists who contributed heavily toward experiments in spectroscopy and astro-photography.

In the last 30 years of the eighteenth century, and into the nineteenth century, the following names dominate the history of astronomy: Simon Newcomb (1835-1909), an American astronomer who completed important studies of the orbits of Mercury, Venus, Mars, Earth, Uranus, and Neptune and developed formulas for explaining the motions of the Moon; Giovanni Schiaparelli (1835-1910), the Italian astronomer, who argued that meteor swarms follow cometary orbits, that Mercury and Venus both rotate on their axis, and that there were lines on the surface of Mars that could very well be caused by canals; and Agnes Mary Clarke (1842-1907), the British astronomer noted for her very popular works on astronomy, including *A Popular History of Astronomy During the 19th Century* (1885) and *Problems in Astrophysics* (1930).

Probably the most important contributors to astronomy in the last half of the eighteenth century, however, were Gustav Robert Kirchhoff (1824-1887), Robert Wilhelm Bunsen (1811-1899), William Huggins (1824-1910), and George Ellery Hale (1868-1938).

Kirchhoff and Bunsen were colleagues who might first be remembered for their discovery of the elements cesium and rubidium. But they were also responsible for some very important work in spectroscopy. Kirchhoff was a German physicist; he was a professor at Breslau from 1850, at Heidelberg from 1854, and at Berlin from 1875. Bunsen, also a German scientist, was on the faculty at Heidelberg from 1852-1859. (His name is especially remembered for an electric cell he discovered—the Bunsen cell.) Using a spectroscope for extensive spectral analysis, Kirchhoff and Bunsen not only discovered the two elements but they also proved that there is a corresponding spectrum for each and every chemical element. They had set the stage for reading stars by the light they give off. A little more than two decades after their breakthrough in spectral analysis, the English astronomer, William Huggins (1824-1910) pioneered the use of spectroscopic telescopes and photographic instruments to determine the chemical makeup of stars; and in 1892, Hermann Vogel (1841-1907) used a spectroscope to determine the radial velocities of stars.

George Ellery Hale (1868-1938) was the American astronomer who organized the Mt. Palomar, Yerkes, and Mt. Wilson observatories. He served as director at Yerkes from 1895-1905 and director at Mt. Wilson from 1904-1913. Hale invented the spectroheliograph, although credit for the invention is also given to Henri Alexander Deslandred (1853-1948) who worked independently of Hale and developed his own spectroheliograph at about the same time. The invention combines the idea behind the heliostat (which is used for reflecting light beams to some distant target) and the spectroscope (which produces spectra for spectrum analysis). It is used primarily for the study of the Sun's atmospheric levels, and what astronomers learn about the Sun helps them to better understand stellar composition and evolution.

## THE MODERN PERIOD

There is no sharp line which indicates the be-

**Table 1-9. Modern Astronomy.**

| Name | Comments |
|---|---|
| Herman Mankowski (1864-1909) | A mathematician rather than an astronomer, his ideas had their influence on astronomy and cosmology; his model of a four-dimensional universe was one of the inspirations for Einstein's relativity theories; made a bold and exciting break with Newtonian physics. |
| Karl Schwartzchild (1873-1916) | Conducted the most comprehensive stellar luminosity surveys ever attempted by a single individual; established the international color index. |
| Eynor Hertzsprung (1873-1967) | Discovered that similar spectral classification does not mean similar absolute magnitude; he demonstrated how absolute magnitudes can be determined by stellar spectra. |
| Victor Hess (1883-1964) | Noted for atmospheric studies and discovery of cosmic rays. |
| Henry Russell (1877-1957) | Contributed to theories of stellar evolution; made extensive spectroscopic studies and published the first diagram which defined the relationship between absolute stellar mangnitudes and stellar spectral types. |
| Henrietta Leavitt (1868-1921) | Published a catalog of about 1800 stars in the Magellanic Cloud; developed, along with Shapley, the period-luminosity curve for Cepheid Variables. |
| Walter Sydney Adams (1876-1956) | Director of Mt. Wilson Observatory; developed a method of spectroscopic parallaxes (along with Ernst Kohlschutter). |
| Arthur Stanley Eddington (1882-1944) | Leading supporter of Einstein's relativity theories; developed the mass-luminosity law and made considerable contributions to theories about the evolution and makeup of stars. |
| Aleksandr Friedmann (1888-1925) | The universe must be either in a phase of expansion or contraction and spectral analyses should reveal the phase through red or blue shift readings of distant galaxies. |
| Georges Lemaitre (1894-1966) | Astrophysicist and mathematician; developed the theory of a universe formed from some primeval atom which contained the "seeds" for everything that now exists. |
| Edwin Hubble (1889-1953) | First to prove that there are numerous galaxies beyond our own; set up a classification scheme for categorizing galaxies. His spectral studies of distant galaxies showed that they are receding at speeds proportional to their distances. |
| Karl Jansky (1905-1950) | Detected radio waves from sources seemingly well-beyond our solar system, thereby founding radio astronomy. |
| Hans Bethe (1906- ) | Among the most famous theoreticians on atomic properties and energy sources; realized that the Sun and stars derive their energy from nuclear reactions. |
| Grote Reber (1911- ) | Verified Jansky's findings and continued to make numerous contributions in the field of radio astronomy. |

ginning of the modern period in astronomy. Many textbooks, in fact, consider the modern period to actually have begun with the early contributions of Nicholas Copernicus in the sixteenth century. Some historians argue that the modern period actually begins with the development of spectroscopy, with Einstein's theory of relativity—or with the space age. For our purposes, however, the modern period in astronomy will be marked as beginning with the work of Herman Minkowski (1864-1909), the Russian mathematician who influenced the relativity theories of Albert Einstein (1879-1955). Minkowski developed a theory that the universe was more than three-dimensional. He considered space and time in terms of a four dimensional continuum. His was a bold and exciting break with Newtonian physics in which only spatial quantities were necessary to identify where and when an event took place. For instance, in Newtonian physics, to know the location of a crash between spacecraft, just knowing longitude, latitude, and altitude would be sufficient; in Minkowski physics, a fourth quantity was necessary: *time*.

Inspired by the ideas of Minkowski, Einstein published in 1905 a paper titled *On the Electrodynamics of Moving Bodies*, which was his *special* theory of relativity; later, in his *general* theory of relativity (1915), he developed his concept of space-time. The general theory's treatment of time, space and motion and how all three are affected by gravity have had a tremendous influence on cosmology. Einstein developed Minkowski's ideas in such a way that he was able to argue convincingly that space and time are not merely abstract terms but, instead, are actual properties. Astronomers and physicists now accept space to be a medium of sorts by which celestial objects are connected and through which we are able to move from one place to the next, much the same as water serves as a medium for the inhabitants of a river or ocean, for it holds things within its realm and allows passage from one point or object to another. As the universe expands or contracts, physicists tell us, space itself expands and contracts. But though space and time are extensions of each other, they are also distinguishable. This means that in modern day physics, there can be no such thing as universal time, for time events can only be measured when spatial relationships are also taken into account.

Between Einstein's 1905 publication of the special theory and his 1915 publication of the general theory, there were a number of technical contributions to astronomy, among them those of Karl Schwartzchild, Eynor Hertzsprung, Victor Hess, Henry Russell, Henrietta Leavitt, Harlow Shapley, Walter Sydney Adams, and Ernst Kohlschutter.

Schwarzchild (1873-1916) contributed significantly to an understanding of the relationship between stellar luminosities and photographic techniques, conducting in 1910 the most comprehensive stellar luminosity surveys ever attempted.[28] He was also indirectly responsible for the international color index established in 1910, for the index was derived from his definition of color given a decade before.[29]

Hertzsprung (1873-1967) was a researcher concerned with the relationships between the absolute magnitude of a star and its spectral type. One of his findings was that despite the fact that certain stars had the very same spectral classification, they had astonishingly varying absolute magnitudes. His work paved the way for the work of Walter Sydney Adams (1876-1956) and Ernst Kohlschutter (1870-1942), who showed how astronomers can determine absolute magnitudes from stellar spectra.[30]

Hess (1883-1964) was an Austrian physicist who later became a U.S. citizen. He was a professor of physics at Fordham University. Hess conducted atmospheric studies with instrument-carrying balloons. In 1912, one of his flights led to the discovery of cosmic rays, a discovery which won him the privilege of sharing the 1936 Nobel Prize in Physics.

Russell (1877-1957) was a Princeton graduate who, after some time in Europe, returned to his alma mater to become a professor of astronomy (1911-1927), research professor (1927-1947) and director of the observatory (1912-1947). In 1947 he joined Harvard University as a research associate. His main contribution to astronomy was his theory of stellar evolution, a theory developed as a result of his extensive spectroscopic studies. In 1913 he published the first diagram which defined the relationship between absolute stellar magnitudes and stellar spectral types.

Leavitt (1868-1921) made an impressive number of contributions in her relatively short life. In 1908, she published a catalog of almost 1800 stars in the Magellanic Cloud and observed that there were much longer periods of varying luminosity in brighter stars. In 1914, working along with Harlow Shapley (1895-1972), she developed the period-luminosity curve for Cepheid variables. Shapley, in turn, was able to work out an astonishingly accurate figure for the size of the Milky Way System, and to estimate that the center of the galaxy is thousands of light years away.

Walter Sydney Adams (1876-1956), the Syrian-born American astronomer was director at the Mt. Wilson Observatory and later a research assistant at the Carnegie Institute; he developed, along with Ernst Kohlschutter, a method of spectroscopic parallaxes.

Thus, in the first 15 years of the twentieth century, a new cosmology was emerging as well as new disciplines within the field of astronomy. In cosmology, the Minkowski universe had given us a four-dimensional model of the cosmos, and Einstein gave us the concept of space-time and the way in which it is influenced by gravity. In astronomy, new strides were taken in the fields of spectroscopy and studies were begun in cosmic radiation.

In the next 15 years of astronomical research and discovery, the foremost contributions would be made by, again, Harlow Shapley, Aleksandr Friedmann (1888-1925), Arthur Stanley Eddington (1882-1944), Georges Lemaitre (1894-1966), and Edwin Hubble (1889-1953).

Shapley, by this time, had been engaged for many years trying to determine the distances of globular clusters and their period-luminosity relationships. His research led him to assume that our galaxy is a great deal more extensive than anyone had imagined, and, even more astounding, despite its tremendous size, it represented only a local group of stellar members. Then, in 1918, he discovered that the Milky Way was surrounded by many globular clusters which, in effect, formed a halo about the Milky Way.

Friedmann's ideas were of value because they represented the first serious challenge to Einstein's model of a static universe. Einstein's ideas were based on non-Euclidean geometric models originally designed by Bernhard Riemann. In 1917, Einstein had argued that the universe is a closed system and that if light were to originate from any point, given enough time it would return to that same point. Friedmann explained that the idea of a closed universe did not really fit in with Einstein's general theory. The universe, Friedmann explained, must be at any given time in one of two possible phases: that of expansion or that of contraction. This idea necessarily included another and that was that spectral analysis would reveal tell-tale red or blue shifts in readings from distant galaxies, these readings indicating the relative directions of the distant star systems as well as their relative speeds.

Eddington was a British astronomer and physicist who became famous for his research on the evolution, motion and make-up of stars. He was also one of the leading supporters of Einstein's theory of relativity. In 1924 he developed his mass-luminosity law, which describes the relationship of the matter contained in a star and the amount of energy it produces—a relationship which must change according to the evolutionary phase of a star.

Lemaitre was a Belgian astrophysicist and mathematician who developed a theory which explains that our entire universe originated from some primeval atom that, for some unexplained reason, exploded. All of the "seeds" for everything that can ever exist were contained in that atom, which exploded with such force that all its components are still in a phase of expansion. The distant galaxies, as well as the inner galaxies, are speeding out into undefined space even now, and this expansion will continue, possibly, for billions of years. Lemaitre's theory, presented in 1927, has since become known as the "Big Bang" theory and is currently the most popular explanation with astronomers for the creation of the universe, although the theory is currently being re-evaluated.

Hubble was an American astronomer and director of the Mt. Wilson Observatory (Fig. 1-20). His research interests were in extragalactic nebulae. He was the first astronomer to prove that there are numerous galaxies beyond the Milky Way and the first to work out a classification system for categorizing galaxies. In 1929, working with Milton Humason at Mt. Wilson, he realized from spectral studies of the distant galaxies that these star systems were moving away from our own at a speed proportional to their distances; he thereby gave new support to Friedmann's ideas.

For the twenty-eight years from 1931 until the first space probes, perhaps the most important developments in astronomy were Karl Jansky's detection of radio signals in space and Hans Bethe's studies of stellar nuclear reactions.

Jansky (1905-1950) was an American radio engineer involved in researching the reasons for static in radio communications. Little did he know that his efforts would lead to a whole new astronomical science. But in 1931 he detected radio waves from sources well beyond our solar system and thereby

Fig. 1-20. Edwin Hubble (1889-1953) was an American astronomer who served as a director of the Mt. Wilson Observatory. He was the first to prove that there are numerous galaxies beyond the Milky Way Galaxy (Courtesy of AIP Niels Bohr Library).

launched the science of radio astronomy.

Bethe is an American physicist who received the Nobel Prize in 1967 for his research into the origins of solar and stellar energy. As far back as the 1930s, he realized that the Sun and the stars derived their energy from nuclear reactions. From 1943-46, he served as director of theoretical research in physics with the Alamos atomic bomb project. He is hailed as one of the foremost theoreticians on atomic properties and energy sources.

Since the 1950s, the great contributions to the storehouse of knowledge about astronomy have come as a result of the massive team efforts that supported the numerous space probes. While these space probes are often used to gather military information, their primary purpose is scientific and since the very first launchings the goals have been fact-finding missions to the Moon and other planets, experiments in deep-space communication, the dynamics of manned and unmanned space travel, and atmospheric studies.

The major U.S. effort began with the Pioneer launchings in the late 1950s, a series of probes which brought back extensive planetary data. Additionally, there were the Luna Orbiter probes which navigated the Moon and photographed both sides of this barren world which, millions or billions of years ago, was a planet soon to be captured by the Earth's gravitational field. Other U.S. probes included the Mariner missions, designed for planetary exploration; the Viking missions, which sought evidence of biological existences on Mars; and the Voyager missions, which have recently extended our knowledge of Uranus and will soon help write new books about Neptune. The Soviet Union has also launched numerous space probes beginning with Sputnik and continuing with Lunik, Lunas, and Veneras missions—from Lunik on, primarily for photographing the Moon and planets and gathering as much data as possible about the solar system, from the physics of the planets and their satellites to the possibility of life in outer space.

The Soviets have recently accomplished a new first in telescope construction. They recently completed, and have in use, the first optical/radio telescope. They've located it at the All-Union Scientific Research Institute of Radiphysical Measurements; and they have scored a "first" with it. It was the means by which the first radio bursts from a red giant, Eta Gemini, were detected. The great advantage to a combination telescope like this is that it is able to detect optical and radio signals at the same time.

Meanwhile, at the Massachusetts Institute of Technology, plans are under way to hold the first session of courses for the new International Space University (ISU). ISU will have its start as a graduate program of summer courses held each year at various universities. M.I.T. is first on the list with the first program to be offered in 1988. In the future, courses will be held in orbit, a rather appropriate environment for a space school.

# Chapter 2
# The Physics of Light

JUST AS THE HOW'S AND WHY'S OF THE NIGHT SKY must have fascinated primitive peoples, so must have the phenomenon of light. What is it? Where did it come from? How did it come? The Sun brought light, of course, and so did the Moon (reflected, of course, but this the ancients never knew), but these were clearly not the only sources of light. There were the stars, the flashes of lightening during storms, the light from the natural fires that erupted from volcanoes or the earth itself during quakes. Light came from many places. But did it come to the eye or did the eye go to it? This was a question that plagued men and women in primitive and ancient and even medieval times. Did it bring heat or was it created by heat? When primitive peoples finally harnessed fire, they began to become even more curious about light. They knew that wherever there was fire, there was both heat and light—and so they began to guess that the Sun and the stars were burning worlds or else worlds on which the gods were burning their own fires.

## ANCIENT AND CLASSICAL IDEAS

Little is known about ancient investigations into the phenomenon of light. The many civilizations which preceded the coming of the Greek city-states—even the Babylonian and Egyptian civilizations—seemed less concerned with the physics of light than with one of its properties, which was heat. But this was only to be expected, because heat was protection from the cold, the means for cooking food and the property that could destroy enemies.

Science had to wait for the sixth century Greek intellectuals before any serious inquiry into light would be recorded. But here again, as in astronomy, the ideas of the ancient Greeks were more creative than factual, but at least these ideas set a pattern for more serious scientific investigations into the nature of light. But in the beginning of their inquiries, the Greeks were little interested in the actual physics of light; they were much more curious about just how light and vision affected each other.

Table 2-1. Theories of Light.

| Name | Ideas |
|---|---|
| Pythagorians (6th and 5th centuries B.C.) | Light originates in the eye; vision is possible because the eye sends visual rays to an object. |
| Empedocles (495-435) | The eye generates light to an object and reads the reflection. |
| Euclid (C. 300 B.C.) | Continued the Pythagorian ideas about vision being possible because of light rays being emitted to an object from the eye. |
| Democritus of Abdera (460-370) | Light rays are generated by luminous objects; the image carried by a ray of light is dispersed in the air so it can be absorbed by the eye. |
| Plato (427-347) | Vision occurs by the fusion of minute particles originating from both light and the eye. When both the eye and light rays fall on an object, they become part of vision and the brain is able to sense the target in detail. |
| Aristotle (384-322) | The eye cannot generate light; light is an all-encompassing substance; it lights the air to produce daylight and when it is burned away by the sun it leaves us with night. |
| Heron (hero) of Alexander (c.60 A.D.) | Established the law of reflection; angle of incidence is always equal to the angle of reflection. |
| Alhazen (965-1040) | Vision occurs because light from an object finds its way to the eye. |
| Roger Bacon (1214-1294) | Drew heavily from Alhazen's ideas; found the focal length for spherical mirrors by geometric means. |
| Galileo (1564-1642) | Attempted to measure the speed of light but decided after at least one experiment that it probably traveled much too fast to have its speed determined. |
| Willebrord Snell (1591-1626) | Discovered the law of refraction. |
| Rene Descartes (1596-1650) | Light is actually a type of pressure; derived the sine law of refraction independently of Snell. |
| Francesco Grimaldi (1618-1663) | First to observe and define the refraction of light. |
| Christiaan Huygens (1629-1695) | Established the wave theory of light. |
| Isaac Newton (1642-1727) | Established the particle theory of light; rays of light differ in their "disposition" to exhibit one color or another. |
| Olaus Romer (1644-1710) | Established that light travels at a definite speed, which he estimated to be 133 thousand miles per second. |
| James Bradley (1693-1762) | Discovered the aberration of light and determined that light must travel in the neighborhood of 188,000 miles per second. |
| Thomas Young (1773-1829) | Applied the wave theory to the refraction and dispersion of light, thereby giving strong support to Huygens' theory. |
| Armond Fizeau (1819-1896) | First to measure the speed of light accurately (as it travels through air). |
| James Clerk Maxwell (1831-1879) | Established the electromagnetic theory; believed in the necessity of the ether. |
| Edwin William Morley (1838-1923) | First to measure the speed of light as it travels through a magnetic field. |
| William Conrad Rontgen (1845-1923) | Discovered X-rays. |
| Heinrich Rudof Hertz (1857-1894) | Demonstrated that radio waves are long transverse waves that move at the speed of light and which can be refracted, reflected and polarized. |
| Max Planck (1858-1947) | Atoms emit and absorb energy in tiny bundles called quanta. |
| Albert Einstein (1879-1955) | Homogenous light is composed of energy grains and replace old light corpuscles by light quanta (photons). Light has the characteristics of both a particle and a wave. |

The Pythagoreans, ever preoccupied with all aspects of science, taught that light originates in the eye (Fig. 2-1). That is, vision is possible because when the eye is directed to perceive some object or environment, it sends visual rays to the target. As these rays create an unbroken path to the object of vision, the eye is able to perceive the color, size, shape and detail of the object. Like many of the other ideas of the Pythagoreans, those on light influenced Greek thought for the next five hundred years and European and Arabian ideas throughout most of the middle ages.

Empedocles (495-435), the Greek philosopher who taught that earth, fire, water, and air were the fundamental particles of nature, was in part responsible for carrying on the ideas of the Pythagoreans. In Empedocles' explanatory model, the eye generates light to an object and then reads the reflection (Fig. 2-2). In other words, to use a modern analogy, vision and sonar detection work in the same way. As a bat might sense objects by screaming shrill sounds which echo back the location and details of an object, so, according to Empedocles, does the eye envision objects by bouncing light rays from its surface. Perhaps Empedocles got his ideas by experimenting with his voice in grottos or caves, and assuming that the process of sight and hearing must be based on the same fundamental principle, decided that the eye and its rays worked like the ear and voice.

The Pythagorean ideas particularly influenced the Greek mathematician, Euclid, a man of whom very little is known except that he produced a collection of geometrical theories and a few other texts, one of which was titled "Optus," a scientific work on light and vision that continued the Pythagorean fiction of the eye as a source of light. As future Greek educators could now point to such respected thinkers as the Pythagoreans and Euclid as proponents of the theory that the eye generates light, the idea was accepted as fact rather than theory. Even an original thinker such as Plato could not completely break away from this tradition. In his *Timaeus*, he explains that vision occurs by the fusion of minute particles originating not only from light but also from the eye. According to Plato, the particles from the eye flow in some invisible stream that never loses contact with its source, so the connection of the eye to its object of vision is never broken (Fig 2-3). When both the eye rays and the light rays fall on an object, it becomes part of vision and the brain can sense its presence and detail. In Plato's scheme, there are two very distinct types of light, one which has its source inside the eye and the other which is generated by some fundamental force of nature. At night, in the absence of natural light, the rays from the eye are smothered much as a blanket would smother a fire.

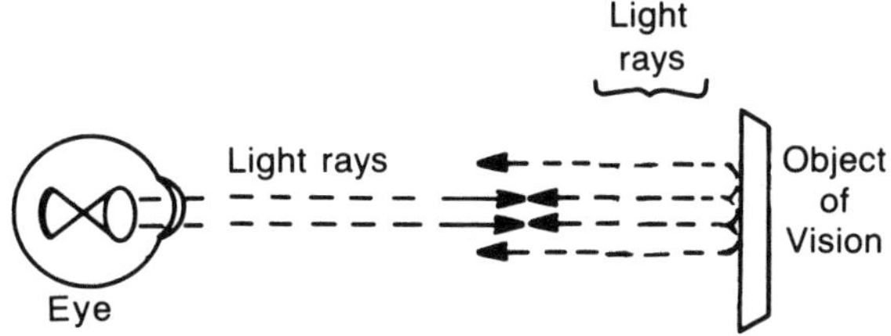


Fig. 2-2. In Empedocle's model, the eye sends light to an object and then reads the reflection.

Democritus of Abdera (460-370), who along with Leucippus, is credited with the idea that all things are composed of imperishable, undetectable things called atoms, taught that there were not two

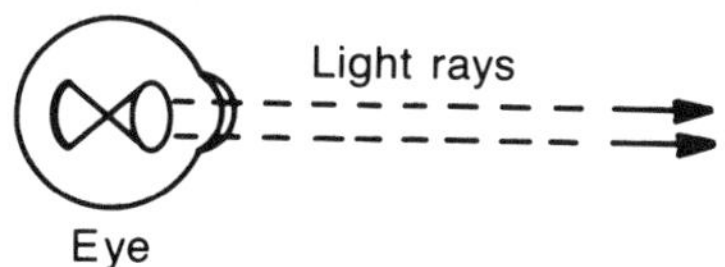


Fig. 2-1. As Pythagorean teachings go, the human eye emits light to objects when directed by the mind to perceive that object.

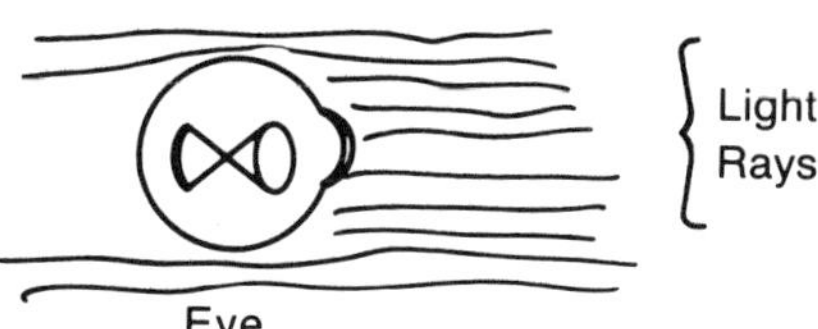


Fig. 2-3. According to Plato (390-310), vision occurs by the fusion of minute particles coming from both the Idea and the object it perceives.

kinds of light rays but only one; and he assumed that light rays were not generated by the eye but rather by luminous objects (Fig 2-4). Once again, Democritus was somewhat ahead of his time; however, he also taught that the image carried by a ray of light was further dispersed in the air so that it could be absorbed by the eye.

Aristotle had very different ideas. He challenged the idea that the eye could generate light; he just could not imagine how it would be possible. But at the same time, he could not understand how light might travel through space to the eye. So, he decided that light must be an all-encompassing substance, perhaps a gas, and in daylight it lit the air. Night comes when the substance or gas is burned away. Colors, he explained, are some extraordinary mix of light and night in the medium between the eye and the object of vision. He noted that vapor causes light to change color and, therefore, clouds change the color of the sun from yellow to red and also bring out the colors found in the rainbow.

Euclid of Alexandria, who flourished in the late fourth and early third centuries B.C. and is remembered mainly for the creation of a number of geometrical theorems, decided he liked Plato's ideas much more than Aristotle's, and carried on the idea that the eye generated light. But Euclid was never really concerned with the nature of light rays; he was much more concerned with the mathematical laws that governed reflection, and he was able to express these laws in purely geometrical terms.

Another way of expressing the law of reflection was given by Heron of Alexandria (d. 62 A.D.), noted scientist and inventor. He explained that light will always take the shortest possible path and therefore the angle of incidence is always equal to the angle of reflection.

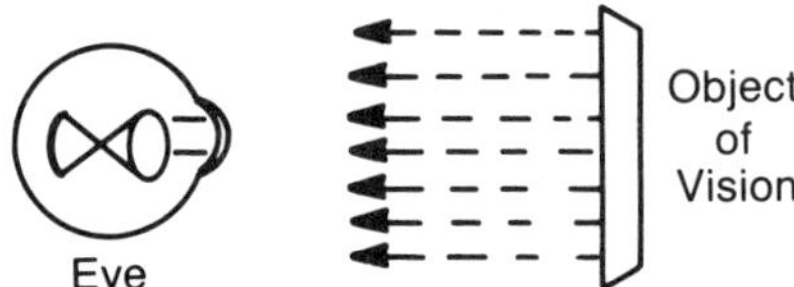


Fig. 2-4. Democritus taught that light rays were generated by luminous objects.

Theories about light did result in some practical applications. Archaeologists have uncovered proof that ancient civilizations had used optical lenses to assist individuals with visual impairments or who needed some tool to help them study very small objects, such as insects. There have been discoveries of glass lenses in coffins found in African sites that date back to the ancient city of Carthage, which flourished until the third century B.C., after which it gradually deteriorated in power and influence and was finally destroyed during the Punic Wars (149-146).

## NEW DIRECTIONS

The Pythagorean idea of the eye as a light source dominated science until well into the eleventh century, during which time an Arabian scientist by the name of Alhazen (965-1040) decided this was highly unlikely. He explained to his colleagues that vision occurred not as the result of the eye sending light to an object but as the result of the light from an object finding its way to the eye. He explained that when light is generated from some source, it runs in each and every direction, filling up all of the space as far as its rays may reach. The same, he said, is true for reflected light. Light coming from a tree or a building which has, as its source, the Sun, will be directed as though it were coming from some source—that is, the rays of light will, in this case also, travel in all directions. Alhazen turned the Pythagorean concept around.

After Alhazen, the greatest contributor to an understanding of the physics of light was Roger Bacon (1214-1294). He drew heavily from Alhazen's ideas and research but went far beyond the latter's investigations.

Bacon was one of the great geniuses in history, a man with a fantastic ability to grasp concepts easily, and a man with a deep interest in philosophy as well as science. He received his higher education at Oxford and at the University of Paris. He believed that philosophy and science were one and the same, but that it is the purpose of science to find the way in which things work and the purpose of philosophy to find the nature and purpose of all things. He echoed the ideas of Anselm and Aquinas when he argued that the search for knowledge had in itself intrinsic value and if science and theology found no coincidence, it was only because man was incapable of seeing it. Despite his genius, Bacon's work was largely ignored after his death. This may have been because in the last years of his life he became obsessed with alchemy and magic and managed to lose his credibility with the scientific community. His work in the area of optics, however, was an exception. He found the focal length for spherical mirrors by geometric means and was able to define the exact point for parabolic mirrors, although he never actually constructed one. He did, however, conduct experiments with magnifying lenses—but never advanced to actually designing anything at all similar to what might pass as a telescope.

In the thirteenth century there were some significant studies on the refraction and reflection of light and these brought about a new understanding of the origin and characteristics of the rainbow. But the real advances in the physics of light came about as astronomers began to realize that without an understanding of the characteristics of light—its dimensions, speed, chemical makeup, etc.—there was little they could learn about the heavens, for, after all, a study of astronomy was fundamentally a study of the lights in the sky.

Perhaps the man most responsible for bringing together the science of astronomy and the physics of light was Johannes Kepler—that is, he at least started the research and interest in merging the two sciences. He realized that vision occurred quite differently than imagined; that is, the eye actually inverts images and does not receive them exactly as they are. In 1604, he published a book on optics that was important for two main reasons. The first was that it marked the beginning of the science of optics. The second was that it inspired the invention of the telescope.

Now, as pointed out in the previous chapter, there is a great deal of argument about who actually did discover the telescope. It may have been a manufacturer of spectacles by the name of Zacharias Jansen; it may have been Hans Lippershey; or it could have been Galileo. It was established by witnesses and documents that Jansen and his son, Hans, made telescopes after 1590 but before 1609. These telescopes are reported to have consisted of one concave and one convex lens and were said to extend a length of about a foot and a half. Lippershey reportedly learned the art of telescope making from the Jansens, and this fact seems to have been common knowledge, for when Lippershey applied for what would be called a patent today, it was denied him. This must have been a startling turn of events for Lippershey for he was a very honest man seeking an opportunity to turn an idea into a fortune. Jansen, however, was not above suspicion, for he once was convicted of counterfeiting for the sole purpose of undermining the Spanish economy. It's possible that Lippershey did learn what he needed to know for that first telescope from the Jansens, but it is probably unlikely that the Jansens had actually constructed a telescope, although what they had was some advanced type of magnifying object. And we know, of course, that though Lippershey did indeed construct a telescope, it was not the kind of instrument that had any application in astronomy—Galileo was the one deserving credit for this kind of instrument. In any event, the important point here is that up until the invention of the telescope, there were really no significant advances in the physics of light or in the kind of technology that might harness light and use it to the advantage of science. A telescope is, of course, an optical instrument, and it owes its development to the advances in optics. But optics, in turn, is dependent on a knowledge of light, for it is a science concerned with the generation and propagation of light, the changes that light effects and undergoes, and the harnessing of light for the advantage of man.

Once Galileo had his telescope, he knew immediately that he had the greatest tool that any astronomer ever had. He must have had a tremendous sense of gratification as he began to make those first discoveries—not simply for the new knowledge he obtained or the fame that these discoveries brought so immediately, but because he had the additional satisfaction of knowing that for the more than 1500 years that men and women were designing and playing with lenses, only he had taken this curious toy and turned it into a powerful scientific instrument. The knowledge that would be gained by its use would forever change man's conception of the universe, alter his philosophy, his theology, his religion.

Once he pointed that telescope to the sky, thereby bringing the members of the galaxy thirty times closer, his discoveries came rapidly: the mountainous surface of the Moon, the extensive star system that the Milky Way actually represented, four satellites of Jupiter (still known as the Galilean satellites).

## THE SPEED OF LIGHT

Just as he was the first to conceive the use of lenses to bring the stars closer to man, Galileo was also the first scientist to attempt to measure the speed of light. It was generally believed in his time that light was an instant phenomena. It just happened, was there. Suddenly there was illumination, and just as suddenly, vision; and there seemed to be absolutely no duration between the two. Galileo, however, sided with a select few who believed that if light did indeed travel in the form of rays, it must move at some finite speed, even if immeasurable. Scientists had already measured the speed of sound at just about 1100 feet per second so there was some support for the idea that light, also, could be measured. But the idea that sound traveled was a much easier concept than the idea that light traveled. The first was certainly obvious. For instance, someone observing a woodcutter chopping away at a distance of several hundred yards would easily note that a second or more might pass between the time the ax hit the tree until the sound of the strike was heard. And sailors engaged in naval battles often saw the flash from a distant cannon before they heard the exploding ammunition. Also, anyone in the vicinity of an erupting volcano would see the eruption before they heard it.

Light, however, seemed so instantaneous. Besides, what could someone measure it against? Well, Galileo had an idea. He decided to conduct an experiment in which lanterns were placed a mile apart and their light was uncovered or covered on signal. When the first lantern was uncovered, the observer at the second lantern would uncover his in reply; Galileo would measure the duration of time existing between originating and verifying signals. It was a rather original idea and one that might have worked except for the extraordinary high speed of light. A ray of light covers a distance of a mile in about one-hundred thousandths of a second; and Galileo would not even be able to recognize that such an interval had taken place, let alone have the means for measuring it. It was no doubt in the back of Galileo's mind that if he could determine the speed of light, he might find an experiment to determine the distances to the Sun, Moon, and planets. But there was little to be gained by his own humble experiment, except that if light traveled at all, it traveled much too fast to have its speed determine.

Actually, the first scientist to discover that light travels at a definite speed was the Danish astronomer, Olaus Romer (also Ole Roemer). Romer (1644-1710) made his observations in 1675 while he was still an assistant at the Royal Observatory in Paris. (He later became a professor of astronomy at the University of Copenhagen as well as royal mathematician.) His ideas on the velocity of light, came as a result of his observations of Jupiter and its satellites. Incentive to use the Jovian satellites in his experiment actually came by accident, for at first he was observing the moons for the sole purpose of verifying or contradicting Galileo's claim that they could be used to measure longitude. Anyway, he knew the millions of miles to the satellites gave him an advantage that he would not have in a terrestrial experiment; so, when he realized that the period of one of the satellites steadily increased over a period of six to seven months, he wondered if he could not use this observation as the basis for determining the speed of light. He guessed that the change in the satellite's period meant a change in the distance that the light rays traveled. A little arithmetic brought him to the conclusion that light traveled at around 133 thousand miles per second.

Romer was off, of course; one, because he needed to know the dimensions of the Earth's orbit for his calculation and these dimensions were not known with accuracy in his time; two, because it was impossible for him to have arrived at the true length of the satellite's period, for astronomy had not yet advanced that far. Still, he was on the right track; his formula was to be respected, the right values just had to be plugged in.

The English astronomer, James Bradley (1693-1762), who discovered the aberration of light, came a little bit closer than Romer in determining the true velocity of light. He came up with a figure of 188,000 miles per second, about 2,000 miles per second too much—at least in terms of the speed of light through air. Actually, the speed of light varies depending upon the density of the medium through which it must move; so, it moves relatively slower through water or glass.

The first scientist to actually measure accurately the speed of light in air was the French physicist, Armond Fizeau (1819-1896). And the first to measure its speed accurately in a magnetic field was Edwin William Morley (1838-1923), an American physicist who was a professor of chemistry at Western reserve (1869-1908).

Morley is also known for his development of the interferometer as an instrument for length and distant measurement, and for his work with Albert Michelson (1852-1931) and D.C. Miller (1866-1941) in measuring the relative motion of the Earth. This last experiment shattered a theory that was thousands of years old. From ancient times right up until Morley's, it was believed that there existed a rarefied element that filled all the upper regions of space. It was called the ether. But in their measurements of the motion of Earth, Morley and company could not find evidence of such an element as the ether. This had important implications in theories about light, for up until that time it was assumed that the ether was absolutely necessary for the propagation of light waves. Thus, the wave theory of light was scuttled and the electromagnetic radiation theory took precedence. (Wave and electromagnetic theories will be discussed again.)

Well, how fast does light travel? In a vacuum, light travels at a speed which has been arrived at by measuring the time it takes a modulated light beam to cover a predetermined distance. The approximate speed is $2.99792458 \times 10^8$m/s $\pm$ 400 m/s. Extended, this means that if you had a vehicle which could travel at the speed of light, in one year you would move 5,880,000,000,000 miles.

## DIFFRACTION AND REFRACTION

For discoveries related to the diffraction and refraction of light, it is necessary to turn the clock back to the sixteenth century. But, first, what is meant by diffraction and refraction?

*Diffraction* is the way in which light, or radiation, will bend around the edge of some obstacle at which it is directed. This bending of light is also affected by a narrow opening. The true cause for the phenomenon is that when light passes any opaque body, it produces a sequence of light and dark lines between the shadowed and luminated areas. Spectroscopic studies are sometimes dependent upon laws of diffraction, because in the diffracting types of spectroscopes, very fine openings are designed into something called a diffraction grating, and the result is a dispersion of light waves into various colors.

*Refraction* means the way in which a light wave is deflected as it passes from one medium to another, as, for instance, when it passes from air into water.

It was Willebrord Snell (1591-1626), the Dutch mathematician who is credited with discovering the law of refraction of light. Snell (Snellius) was a professor of mathematics at the University of Leiden when he formulated the law. He stated that as light travels from one medium to another, the sine of the angle of incidence (the perpendicular line at the boundary between the mediums right at the point of refraction) will always be the ratio of the refracting medium's index of refraction to the first medium's index.

And it was Francesco Grimaldi (1618-1663), the Italian physicist and Jesuit priest, who was the first to describe the fringes bounding the shadow of an object (illuminated by a small ray of light), thereby being the first to observe and define the diffraction of light. But it was the Danish physician and physicist, Erasmus Bartholin, who discovered the phenomenon of double refraction, in which a beam is split into two independent beams by certain crystals, and then refracted at different angles.

## THE PRESSURE THEORY

But, so far, what has been discussed about light are simply discoveries related to its characteristics and not its actual nature. To do this, one might begin with the ideas of Rene Descartes. Now, Descartes (1596-1650) is best remembered for his philosophical discourses and great cosmological argument for the existence of God, but he also made important contributions to science and produced works not only on mathematics and psychology, but also on optics. Descartes was intrigued with the subject of light; it was after all, such a strange occurrence. What could it be? Well, his observations indicated that light was actually a type of pressure, for its properties seemed to follow the same basic laws.

Observing that the laws of reflection could very well be applied to a rebounding ball, he went on to derive the sine law of refraction—independently of Willebrord Snellius, although the law (Snell's law) was named after the latter. Applying the law, Descartes showed that only elliptical or hyberbolical lenses could be used to focus light on any given object; and he used the law to explain why certain aspects of a rainbow could not be seen at specific angles. But Descartes believed that light moved much more swiftly in more dense mediums. Another French researcher, Pierre Fermat (1601-1665), a magistrate who spent his free time working at mathematics, eventually challenged Descartes and showed that it actually moved faster in lighter mediums.

Descartes theory that light was pressure that could be transmitted through transparent media dominated the physics of light until the late 1600's when Isaac Newton's ideas began to attract attention. Newton (1642-1727), considered to be the greatest scientist to have ever lived, was a mathematician and physicist. He discovered the law of gravitation and developed calculus, was a professor of mathematics at Cambridge University from 1669-1701, and president of the Royal Society from 1703 until his death. In 1704, he published his *Opticks*, in which he argued for the particle theory of light (Fig. 2-5).

## THE PARTICLE THEORY

"Light is not homogeneal," Newton wrote, "but consists of different rays, some of which are more refrangible than others."[1] These rays of light, he went on to explain, also differed in their "disposition to exhibit this or that particular color."[2] It was this fact of color, as well as the fact that light moved in straight lines, that led Newton to develop his particle theory. According to him, what we perceive as one indivisible flash, or steady pressure of light, is actually a series of very minute particles traveling rectilinearly at an extreme velocity.

By Newton's time, the reflection and the refraction of light were already well-studied. Additionally, thanks to Romer's experiments, it was already accepted that light traveled at some finite speed. All but refraction were typical of particle streams, so all Newton had to do was illustrate that refraction would occur as the speed of light gained while the density of the medium through which it passed increased.[3] This he proved, now upsetting Fermat and supporting Descartes.

## THE WAVE THEORY

Christiaan Huygens (1629-1695), the Dutch mathematician and physicist whose contributions to astronomy were discussed in the previous chapter, had argued for the wave theory of light; after his death, Newton's ideas took the foreground. But Huygens had already looked at most of the arguments for the particle theory and since discarded them. He saw some particular difficulties in accepting that the source of any light could so fill the air in every direction that was not blocked if it were emitting distinct particles. Additionally, it was quite easy to demonstrate that beams of light coming from separate directions could cross each other without in anyway interfering with each other. Now, if light were indeed simply consisting of particle streams, when rays of light crossed each other, there should be some level of chaos, with beams going one way or another and lighting objects which were not in the direct line of either beam. Huygens, therefore,

Fig. 2-5. Isaac Newton (1642-1727) developed the particle theory of light. He explained that light consisted of different rays, some more refrangible than others (Courtesy of Bausch & Lomb Optical Co. and AIP Niels Bohr Library).

decided that the particle theory was invalid, and in its place he substituted the wave theory.

For centuries, physicists went one way or the other and then back again; there were times when Newton's ideas were supported with such numbers that Huygens' wave theory was only discussed in private; then there were times when the scientific community became completely frustrated with their attempts to use Newton's ideas to account for things like interference or diffraction and began to look again at Huygens' theories.

In the early part of the nineteenth century, the tide of research and opinion began to lean once and for all to the wave theory. Beginning in 1802, Thomas Young (1773-1829), the English physicist, physician and Egyptologist, presented a series of papers which brought new validity to Huygens' arguments. Most importantly, Young was able to apply the wave theory to the refraction and dispersion of light, applications important to acceptance of the theory by the scientific community. Additionally, the work of Augustin Jean Fresnel (1788-1827), the French physicist and engineer, Gustav Robert Kirchhoff (1824-1887), the German physicist, and others established the transverse-wave theory, which established once and for all that light was a wave vibration.

The final, and successful, challenge to Newton's particle theory actually came in the mid-nineteenth century when it was found that the speed of light varied in different media in a way completely opposite to what Newton had predicted. If light was a wave form, through what medium did it travel? The curious aspect of light was that it could actually move through a vacuum. Not only could it move through a vacuum at an incredible speed, but that speed was even greater than the one in which it moved through a physical medium like air or water. Physicists, therefore, accepted the idea that there must be some cosmic medium through which light moved. This universal medium was held to be some type of elastic solid. This allowed for the phenomenon of light traveling through a vacuum; the only problem was that there was no scientific evidence of the ether.

James Clerk Maxwell (1831-1879), the Scottish physicist who became the first professor of experimental physics at Cambridge University, turned back to ancient ideas and re-sparked the concept of something called the *ether*, which he said existed throughout space and transmitted electromagnetic vibrations. During his life, and to the end of the nineteenth century, many experiments were tried to confirm the existence of the ether but they all failed. It was finally, the Michelson-Morley experiments that laid to rest theories about the existence of the ether as a medium for transmitting electromagnetic radiation.[4]

Despite the fact that Maxwell was wrong about the existence of the ether, his electromagnetic theory was mathematically sound. From it, predictions were made about the existence of electromagnetic waves that traveled at a constant speed in a vacuum; and later experiments in Germany and England confirmed Maxwell's theories. There was now an established connection between light and electromagnetism.

Light is some kind of hybrid in the world of physics, and this was soon realized by the results of experiments by Heinrich Rudolf Hertz (1857-1894) and William Conrad Röntgen (1845-1923).

Hertz was a German physicist most famous for his study of radio waves which he demonstrated to be long transverse waves that moved at the speed of light, and which behaved very much like light in that they could be refracted, reflected and polarized. While he did indeed confirm Maxwell's theories, he did upset one of Maxwell's conclusions. Maxwell had implied that the energy of electrons emitted as the result of ultraviolet light on some metallic surface, would be dependent upon the intensity of the radiation. Hertz, on the other hand, found this to be untrue. According to his research, that energy would be dependent upon the frequency of the radiation.

Röntgen, another German physicist (Fig. 2-6), is remembered mainly for his discovery of a short-wave ray named after him but known today as the X-ray. His research and discoveries brought a new dilemma to physicists interested in a formidable theory of light. The X-rays with which he worked exhibited the same wave phenomena as light, for they were, in fact, forms of electromagnetic energy just

Fig. 2-6. William Röntgen (1845-1923) discovered X-rays (Courtesy of AIP Niels Bohr Library).

as was light. But they also acted very much like particles. Other research about the time of Röntgen's, which centered on the way in which energy was emitted and absorbed, presented additional problems, for this research indicated that light must have corpuscular characteristics.

So, with the advent of the twentieth century, science was not sure in which way it must describe light, for many of its characteristics could be explained in terms of a corpuscular (or, particle) theory and others only in terms of a wave theory. Light often acted as though it consisted of particles; and light often acted as though it consisted of waves.

## THE QUANTUM THEORY

In physics, there is something which is called a blackbody. This is an ideal energy radiator, a substance that absorbs all of the radiant energy that comes to it, and reflects none of that energy. Carbon is one example of a blackbody and lampblack another, although in the case of lampblack there is a 1 to 2 percent reflection. Now, in the latter part of the nineteenth century, there were some problems related to blackbody radiation that continued to plague scientists and added to additional confusion about the nature of light. Blackbody radiation refers to the tendency of an object to change color as its temperature becomes more and more intense. Eventually no more changes in color will be observed, for as the temperature really becomes intense the new radiation given off is invisible, as in the case of ultraviolet radiation.

That objects would change color as they were heated was no new discovery by physicists. This was something that had been clearly observed in ancient times when metal was heated for shaping or when wood fires gave off colored flames as they grew hotter then cooled. There were always close associations between heat and light and its colors. But what had physicists particularly perplexed was the way in which heat became light. Where was the connection? How did the transition occur? If they could not explain it mathematically, they could not understand it.

The answers began to come with the theoretical work of the German physicist, Max Planck (1858-1947). He discovered, through a tedious process of developing formulas by a trial-and-error process, that atoms emit and absorb energy not in continuous streams as the classical physicists believed but in tiny bundles called *quanta*. But his findings at first seemed to be a contradiction, for light was energy and it was now almost fully-accepted that light was a waveform; how, then, could his research conclude that light would be emitted as quanta rather than as waves? Despite the fact that his work, after being checked and rechecked, withstood all challenge, Planck could not believe the results himself, and cared little whether or not the scientific community accepted his findings.

Confirmation of his work eventually came as the result of research by Albert Einstein (1879-1955), the American theoretical physicist. German born, he was professor of physics and director of theoretical physics at Berlin's Kaiser Wilhelm Institute. Problems with the Nazi government forced him to revoke his German citizenship and make his way to the U.S. where he joined the staff at the Institute for Advanced Study in Princeton (1933-1945).

In his *The Evolution of Physics*, co-authored with Leopold Infeld, Einstein explained that there were experimental results that could not be predicted by the wave theory.[5] "Let us be deliberately unjust to the wave theory of light," he wrote, "forgetting its great achievements, its splendid explanation of the bending of light around very small obstacles. Without attention on the photoelectric effect, let us demand from the theory an adequate explanation of this effect."[6]

He went on to explain that it was impossible to use the wave theory to explain the independence of the energy of electrons from the intensity of the light by which they were emitted from a metal plate, a point which was made earlier in this chapter. He, therefore, suggested that another theory be considered, and he directed his readers back to Newton's corpuscular theory. "To keep the principal idea of Newton's theory," he wrote, "we must assume that homogeneous light is composed of energy-grains and replace the old light corpuscles by light quanta, which we call *photons*, small portions of energy, traveling through empty space with the velocity of light. The revival of Newton's theory in this new form leads to the *quantum theory of light*. Not only matter and electric charge, but also energy of radiation has a granular structure; i.e., is built up of light quanta. In addition to quanta of matter and quanta of electricity there are also quanta of energy."[7]

The quantum theory sufficiently explains the photoelectric effect: photons fall on a metal surface; the photons interact with the atoms in such a way that electrons are freed, with each extracted electron having a level of kinetic energy equivalent to the energy of the photons minus the energy spent to free the electrons. But Einstein knew the scientific community would be perplexed by his findings, for they would have to readily admit that whatever theory of light they must now formulate, it must be a synthesis of the wave and particle theories. As Einstein pointed out, "There seems to be no likelihood of forming a consistent description of the phenomenon of light by a choice of one of the two possible languages. It seems we must sometimes use the one theory and sometimes the other, while at times we may use either."[8]

In the years that have passed since Einstein's first announcements of his research, experiments have continued to confirm his ideas. Light remains a conundrum to the scientific community. Today they no longer refer to light as a wave or a particle, but as a photon. Light is actually made up of something still very strange to the scientific world. It resembles waves in some ways, particles in another, but it is neither, or it is simultaneously both.

## LIGHT AND COLOR

The first observations of light and color that have even the slightest scientific bearing were probably those done by the Roman philosopher and statesman, Seneca, who died in 65 A.D. at about the age of 70, a suicide committed in response to accusations that he took part in a political conspiracy. Seneca's contributions to moral and political philosophy and to literature in general are extensive. His adventures in science were mainly exercises in logic or else occasional attempts to find the fundamental laws that might support his philosophy or explain some natural phenomena that caught his passing interest. Color always caught his imagination. Just how did it occur, he often asked himself, that some things are one color and others are another? And why did rainbows generate so many different colors?

Seneca noticed that the colors of the rainbows were very similar to what one would perceive when light passed through a glass cut with three corners and having a sharp edge. He, therefore, decided that the eye was indeed a very deceptive organ and one that could not be relied upon, for how could it see color where there was indeed no color—unless, of course, color was an intrinsic ingredient of light, becoming apparent when light interacted with certain objects. Seneca concluded that there were at least two different laws affecting color, one which caused the colors of the rainbow, the other which caused color in certain objects.

Modern ideas about light actually owe their origins in the work of that mastermind, Isaac Newton. Newton's experiments led him to theorize that as the rays of light differ in degrees of refrangibility (ability to be refracted), so do they differ in degrees of color. "Colors are not qualifications of light," he explained, "derived from refractions, or reflections of natural bodies (as is generally believed), but from original and connate properties...Some rays are disposed to exhibit a red color and no other, some a green and no other, and so on. Nor are there only rays proper and particular to the more eminent colors, but even to all intermediate gradations."[11 9]

Newton went on to explain that there was a direct relationship between refrangibility and color. Refrangibility refers, of course, to the way in which light is deflected from its normal path when it moves through any medium. As Newton explained it, "to the same degree of refrangibility ever belongs the same color, and to the same color ever belongs the same degree of refrangibility. The least refrangible rays are all disposed to exhibit a red color, and contrarily those rays, which are disposed to exhibit a red color, are all the least refrangible: So the most refrangible rays are all disposed to exhibit a deep *Violet Color*, and contrarily those which are apt to exhibit such a Violet Color, are all the most refrangible. And so to all the intermediate colors in a continued series belong intermediate degrees of refrangibility.''[10]

Newton was the first to see color as a natural phenomenon that could be investigated and explained. Before this time, color was simply accepted to be a characteristic of the world occurring probably by accident or else it was conceived to be some mystical phenomenon beyond the bounds of man and his understanding. Newton, however, conceived of God as a supreme scientist who not only created the universe but designed it according to strict mathematical and physical laws. He felt that the search for an understanding of the universe was a search for God himself. He was a natural philosopher and natural theologian who believed that it was within the intellectual power of man to understand the matter and systems of the world.

One of his most important experiments with light consisted of a windowshade, a triangular glass prism and a screen. His purpose here was to bend light, note the colors that were created by the different refracted rays, and note the relationships, the results of which were quoted in the above excerpts from his writings. When the light passed through the glass prism, its beam was refracted and spread and the result was a projection of seven different colors on his screen. These were violet by the most refracted beam, blue by the second-most, greenish-blue by the third-most, green by the fourth-most, yellow by the fifth-most, orange by the sixth-most and red by the least refracted.

Newton next isolated one of the refracted beams so that it would pass through a second glass prism. His intention here was to determine whether or not the prisms themselves might be affecting the color of the refracted light. The result was that there was no change in color as the refracted light passed onto the screen.

''The species of color, and degree of refrangibility proper to any particular sort of rays, is not mutable by Refraction, nor by reflection from natural bodies, nor by any other cause, that I could yet observe,'' he argued. ''When any sort of rays have been well parted from those of other kinds, it hath afterwards obstinately retained its color, notwithstanding my utmost endeavors to change it.''[11]

He refracted the source beam with prisms, reflected it with all kinds of objects, intercepted it with different colored devices—in short, tried to affect the beam in every possible way in order to change the range of colors which he was able to produce, but never could. He knew by mixing different rays he could produce even more colors. However, he cautioned that ''transmutations made by the convening of diverse colors are not real; for when the different rays are again severed, they will again exhibit the very same colors.'' He gave as example the colors blue and yellow, which, when mixed, produced green.

''There are, therefore, two sorts of colors,'' he continued. ''The one original and simple, the other compounded of these. The original or primary colors are, *Red, Yellow, Green, Blue* and a *Violet-Purple*, together with *Orange, Indigo*, and an indefinite variety of intermediate gradations.''

It is important to note here that Newton's experiments defined the physical property of light that produces color; that is, it is the refraction of light that separates a beam to reflect the different colors of the spectrum. Newton's experiments did not reveal anything about the perception we have of colors, nor did Newton ever try to insinuate that they did. A beam of light may separate into seven primary colors, but the actual way in which these colors may be perceived by an individual is another matter entirely.

Newton's ideas about color were never received with great enthusiasm. This may possibly have been because there were traditional attitudes toward the subject that were not easily broken, one of these being the idea that light was a fundamental entity, could not be broken down or divided in any way. Remember, in Newton's time, there was still no accepted theory about exactly what light was. Newton was advocating the particle (or corpuscular) theory, and Huygens was finalizing his ideas on a wave theory.

It was not until the early 1800's that Newton's ideas about color began to be re-evaluated. This re-

**Table 2-2. Colors of the Spectrum.**

| Color | Wavelength (nanometers) |
|---|---|
| Violet | 380-435 |
| Blue | 435-500 |
| Cyan (Greenish-Blue) | 500-510 |
| Green | 510-555 |
| Yellow | 555-590 |
| Orange | 590-625 |
| Red | 625-740 |

Cyan is included here, but in many cases only the other six monochromatic colors are considered a part of the spectrum. Of these colors, generally only red, green and blue are considered the primary colors.

evaluation came as the result of the work of Thomas Young (1773-1829), one of the scientists who regenerated interest in the wave theory. He conducted a series of experiments which illustrated that an extensive number of hybrid colors can be produced by mixing red, violet and green lights in certain proportions. This brought him to conclude that the human eye has three different color receptors.

James Clerk Maxwell tells us that since Young, "by one of those bold assumptions which sometimes express the result of speculation better than any cautious trains of reasoning," had determined that these three distinct means of color sensation in the retina were the reason for the range of color perception, and these sensations must also be produced in different degrees by the differing rays.[13]

In Young's scenario, each nerve acts, not by informing the mind of the length of the undulation of a beam, but simply by being more or less affected by the rays which fall on it. The sensation caused in each nerve can detect no other change than that of increase or decrease, and the nerves which correspond to the red sensation are affected chiefly by the red rays but also by the rays which come from other parts of the spectrum, just as a green glass transmits green rays primarily and other rays secondarily.

Maxwell illustrates Young's theory in a paper called "Experiments on Color, As Perceived by the Eye."[14]

"Let a plate of red glass be placed before the camera, and an impression taken. The positive of this will be transparent wherever the red light has been abundant in the landscape, and opaque where it has been wanting. Let it now be put in a magic lantern, along with the red glass, and a red picture will be thrown on the screen.

"Let this operation be repeated with a green and a violet glass, and, by means of three magic lanterns, let the three images be superimposed on the screen. The colour of any point on the screen will then depend on that of the corresponding point of the landscape; and, by properly adjusting the intensities of the lights, etc., a complete copy of the landscape, as far as visible colour is concerned, will be thrown on the screen. The only apparent differences will be, that the copy will be more subdued, or less pure in tint, than the original. Here, however, we have the process performed twice—first on the screen, and then on the retina."

Young's theories were further advanced by Maxwell himself and then by other researchers. The history of research in how the human brain perceives color is extensive and interesting, but for our purposes, we have traced it far enough. What is important at this point is the idea that a beam of light can be spread out and analyzed. This idea has led to the development of a discipline within the field of physics called spectroscopy, which deals with the interpretation of interactions occurring between matter and electromagnetic radiation.

What the telescope is to astronomy, the spectroscope is to spectroscopy. It forms and examines optical spectra. When light is received into the spectroscope, it is spread by a prism. The technique here is little different than that used by Newton in his experimentation with refracted rays. But with the spectroscope, the energy rate for each wavelength can be measured independently. Now this idea has had important application in the field of astronomy where the telescope is used as the light-gathering means. An astronomer points his telescope to some distant light, perhaps that coming from a star. The telescope is wired to a spectrograph which will disperse any electromagnetic radiation detected from the source into a spectrum, and then photograph or map the spectrum. Because different chemical elements give off different kinds of light, and therefore different spectral lines, astronomers can determine the chemical makeup of a star. It is amazing what they are able to determine by spectroscopic studies. They can tell not only the number of elements contained in the star but also their relative quantities; they can determine the star's temperature; they can estimate the intensity of its magnetic field; they can even determine the speed at which a star is moving from or toward the Earth.

## ELECTROMAGNETIC RADIATION

Just what is light? Or, rather, just what does science know about light so far?

Before the question should be answered, some introduction to electromagnetic radiation is necessary. This introduction must necessarily center about something called the electromagnetic spectrum and the types of radiation of which it consists. This is because, fundamentally, light may be considered to be the full range of radiation forms which make up the electromagnetic spectrum, although what we call ''light'' is often a term limited to light that can be perceived by the human eye.

Electromagnetic radiation is that kind of radiation that is transmitted in the form of electromagnetic waves. As the name ''electromagnetic'' would imply, these waves are comprised of both electrical and magnetic fields. The energy produced is actually the result of the acceleration of a charged particle. The types of radiation which are particularly important to astronomers for ''reading'' stars are radio waves, infrared radiation, visible light waves, ultraviolet radiation, X-rays and gamma radiation. Each of these types of radiation carry with them important information about celestial objects. With the right equipment and the know-how, this information can be deciphered.

Generally, for the purposes of this book, all of these different forms of radiation are divided into visible light or invisible light, as all of the forms of radiation applicable to astronomical research may be converted to light images in one way or another. This becomes more evident as we continue. But, for an immediate explanation, consider this. Outside what is termed the ''visible spectrum''—immediately outside in terms of higher and lower wavelengths and frequencies—are infrared and ultraviolet radiation. These may be termed infrared light or ultraviolet light because while we cannot see by infrared or ultraviolet there are insects and animals who can see these forms of radiation, or at least see by them. For instance, many types of snakes are suspected of being able to track their prey at night by using infrared sensors and bumble bees are believed to find their way from flower to flower by ultraviolet sensing capabilities. What we are generally looking at, then, when we look at the electromagnetic spectrum is light in its different forms, or wavelengths.

We know that wherever there is light, there is heat. And heat is a common characteristic of all forms of electromagnetic radiation, at least from a theoretical standpoint. However, in some cases the theory has been turned to actuality, for it is well established that X-rays, ultra-violet, visible and infrared radiation, microwaves and radiowaves can be detected by heat-sensing instruments. As in the case

**Table 2-3. Electromagnetic Spectrum (courtesy of McGraw-Hill Encyclopedia of Astronomy).**

| Frequency, Hz | Wavelength,m | Nomenclature | Typical source |
|---|---|---|---|
| $10^{23}$ | $3 \times 10^{-15}$ | Cosmic photons | Astronomical |
| $10^{22}$ | $3 \times 10^{-14}$ | $\gamma$-rays | Radioactive nuclei |
| $10^{21}$ | $3 \times 10^{-13}$ | $\gamma$-rays, X-rays | |
| $10^{20}$ | $3 \times 10^{-12}$ | X-rays<br>Positron-electron annihilation | Atomic inner shell |
| $10^{19}$ | $3 \times 10^{-11}$ | Soft X-rays | Electron impact on a solid |
| $10^{18}$ | $3 \times 10^{-10}$ | Ultraviolet, X-rays | Atoms in sparks |
| $10^{17}$ | $3 \times 10^{-9}$ | Ultraviolet | Atoms in sparks and arcs |
| $10^{16}$ | $3 \times 10^{-8}$ | Ultraviolet | Atoms in sparks and arcs |
| $10^{15}$ | $3 \times 10^{-7}$ | Visible spectrum | Atoms, hot bodies, molecules |
| $10^{14}$ | $3 \times 10^{-6}$ | Infrared | Hot bodies, molecules |
| $10^{13}$ | $3 \times 10^{-5}$ | Infrared | Hot bodies, molecules |
| $10^{12}$ | $3 \times 10^{-4}$ | Far-infrared | Hot bodies, molecules |
| $10^{11}$ | $3 \times 10^{-3}$ | Microwaves | Electronic devices |
| $10^{10}$ | $3 \times 10^{-2}$ | Microwaves, radar | Eelctronic devices |
| $10^{9}$ | $3 \times 10^{-1}$ | Radar<br>Interstellar hydrogen | Electronic devices |
| $10^{8}$ | 3 | Television, FM radio | Electronic devices |
| $10^{7}$ | 30 | Short-wave radio | Electronic devices |
| $10^{6}$ | 300 | AM radio | Electronic devices |
| $10^{5}$ | 3000 | Long-wave radio | Electronic devices |
| $10^{4}$ | $3 \times 10^{4}$ | Induction heating | Electronic devices |
| $10^{3}$ | $3 \times 10^{5}$ | | Electronic devices |
| 100 | $3 \times 10^{6}$ | Power | Rotating machinery |
| 10 | $3 \times 10^{7}$ | Power | Rotating machinery |
| 1 | $3 \times 10^{8}$ | | Commutated direct current |
| 0 | Infinity | Direct current | Batteries |

From McGraw-Hill Encyclopedia of Astronomy (courtesy McGraw-Hill).

of visible light, the individual quantum for all forms of electromagnetic radiation is the *light quantum*—or, as it is better known, the *photon*.

The forms of electromagnetic radiation with which we are primarily concerned are those with the shorter wavelengths and higher frequencies, beginning with far-infrared and ending with gamma (y-rays) rays. There are wide intervals in the range of frequencies for the different forms of electromagnetic radiation so there is a great deal of overlapping in terms of specific characteristics related either to the detection or the generation of these waves.

Gamma (y-rays) radiation is possibly the most intense form of electromagnetic radiation, basically being highly energized X-rays with very short wave lengths. It is one of the three forms of natural radioactivity, the others being the alpha particle and the beta particle. Radioactivity results from the disintegration is usually the form of radiation that we are discussing here, electromagnetic. Gamma-ray astronomy is still in its infancy but, as you will see later on, it is being used and may have exceptional potential for helping physicists and astronomers study the universe.

X-rays also have very short wavelengths and are invisible just like gamma rays, although the wavelengths for the X-rays still are not quite as short (nor the frequency as high) as that of gamma rays. Both X-rays and gamma-rays can be highly destructive, but fortunately these forms of radiation are absorbed in the Earth's atmosphere long before they reach the surface of the planet. That means both X-ray and gamma-ray astronomy cannot be conducted from the great observatories; astronomers must reach into and above Earth's atmosphere with their detecting instruments to collect the data they need for analyses. Like gamma-ray astronomy, X-ray astronomy is also still in its infancy, but it has helped to increase knowledge about stars, cosmic gas and may be an inroad to further understanding about black holes or pulsars.

As we continue to scan the electromagnetic spectrum, moving from the shorter wavelengths characteristic of gamma and X-rays (Fig. 2-7), we come into the ultraviolet range of the spectrum. Here, we remain in the invisible range of the electromagnetic spectrum. But while we cannot see ultraviolet radiation, we can sometimes see the results of its presence. For instance, when our bodies are tanned by the Sun, it is the ultraviolet rays coming from this star that have caused the browning of our skins. Astronomers zero in on ultraviolet radiation as a means of detecting the hottest stars, as well as for gathering information about the great clouds of gas that exist between stars or systems of stars.

Once out of the ultraviolet range, we come to what physicists define as *visual light*, or the visible part of the electromagnetic spectrum. The wavelengths for visible light lie almost at the center of the electromagnetic spectrum. Visible light consists of relatively short waves, though not as short as gamma, x and ultraviolet waves. As Isaac Newton quite aptly demonstrated, the different wavelengths within the range of visible radiation can be perceived in different colors, those colors that we find in the rainbow. The longest waves of visible radiation are perceived as red light and the shortest as blue light. As different chemicals will produce different colors of light, astronomers can determine the chemical stars by recognizing the different colors of light that they generate.

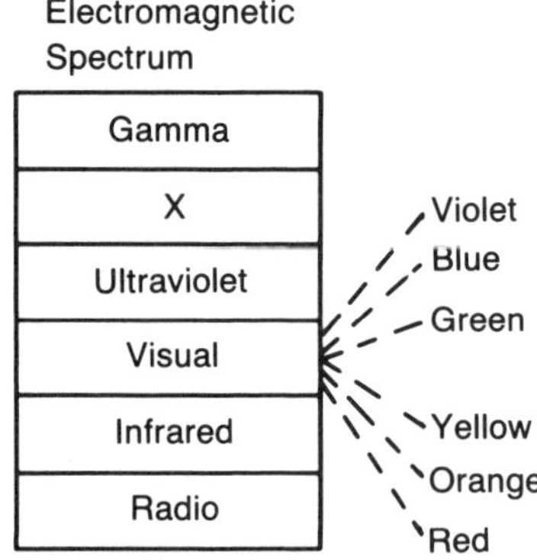

Fig. 2-7. A block diagram of the electromagnetic spectrum and the wavelength ranges for the different colors.

Now, as we continue to scan the electromagnetic spectrum, we cross the range of visible light to a form of radiation with a wavelength in excess of radiation in the visible ranges. This is infrared radiation. Infrared radiation is basically heat radiation, although technically any type of radiation must generate some type of heat. However, when astronomers are able to detect infrared radiation coming from some celestial object, they know that the source is a relatively cool world, or piece of world. Thus, when they turn their infrared telescopes to one of the other planets in our solar system, they expect to find a great deal of infrared radiation, for, because of their relatively low temperatures (less than 100° K), planets will always radiate in the infrared range of the electromagnetic spectrum.

The majority of astronomical objects, especially those which consist of solid matter having lower temperatures, radiate most of their energy at the infrared wavelengths.[15] Additionally, there are certain events in the evolutionary cycle of an astronomical object, including starbirth and the condensation of planetary systems, that can best be investigated only through infrared astronomy.[16] The problem to astronomers, however, is that the atmosphere of the Earth blocks almost all of the infrared radiation coming into it and constructing astronomical instruments that can pick up infrared rays is no easy technological feat. Thus, it is a frustrating fact that most of the infrared signals coming from outer space remain undetected. This means that a large part of the sky remains no more than a blank page as far as invisible astronomers are concerned.

Continuing to the next range in the electromagnetic spectrum—still moving, of course, from the smaller to the larger wavelengths while moving from higher to lower frequencies—we come to the radio waves. Radio radiation actually covers a huge spanse of the electromagnetic spectrum, and is generally expected to consist of wavelengths of one millimeter and greater.

Radio astronomers have contributed greatly to developing a map of the sky, assisting their colleagues in other branches of astronomy in understanding what and where lie the many unseen objects in the sky. Nature presents special problems to radio astronomers as it does to astronomers working at other wavelengths. Many of the radio signals detected in the sky cannot be traced to any source. The complexity of the universe, its mysterious geometry, and, perhaps, the unknown forces which work within and on it may be creating radio illusions as easily as they create optical illusions.

## PROPERTIES AND NATURE OF LIGHT: A SUMMARY

With some understanding of the electromagnetic spectrum, it is possible to better understand what is known about that strange phenomenon called visible light. Let's put together the facts that may be compiled from the history presented earlier.

When physicists refer to visible light they refer to that part of the electromagnetic spectrum that can be seen. In a much wider perspective, light is sometimes considered to be the entire electromagnetic spectrum, visible or invisible, depending upon the range of wavelengths by which it is encountered at any given time. In this text, we generally treat light as the full range of radiation making up the electromagnetic spectrum, and thus the title of this book: *Studies in Starlight*. Astronomers have had to harness a number of different types of technology to detect light at its different wavelengths and have developed these brand new astronomies: optical astronomy, infrared astronomy, ultraviolet astronomy, X-ray astronomy, gamma ray astronomy and radio astronomy. Optical astronomy remains highly dependent upon the great telescopes at the great observatories located around the world, and the painstaking and systematic observations of teams of astronomers, but the other types of astronomy need radio telescopes to detect radiation at wavelengths greater than one millimeter, special photographic plates to detect infrared or X-rays, and many other ingenious systems which allow them to detect other types of electromagnetic radiation.

No matter what its source, light is transmitted in straight lines but begins to expand immediately after origination and continues to cover a wider and wider area as it travels. Physicists can tell us that the speed of light is almost exactly 186,282.396 miles per second, and knowing the speed of light helps astronomers gauge the distance to other parts of the solar system and galaxy, although distance measurement techniques used by astronomers will vary considerably, depending upon whether they are measuring distances within or without the galaxy and the type of celestial object (such as double stars) to which they wish measurement.

When light meets an object in its path, it may be reflected, absorbed, or just scattered in every possible direction. If the object has a smooth surface, the light will be reflected (Fig. 2-8); if it has a rough surface, then absorption or scattering takes place. The whiter the surface which the light strikes, the more equal will be the scattering of the waves. The darker the surface, the more varying the distribution during scattering. A perfectly black surface should, theoretically, absorb all the light which it receives.

The most observable property of light is its reflection. Smooth surfaces will reflect it in such a way that its angles of incidence and reflection will be the same. The less smooth surfaces, as already

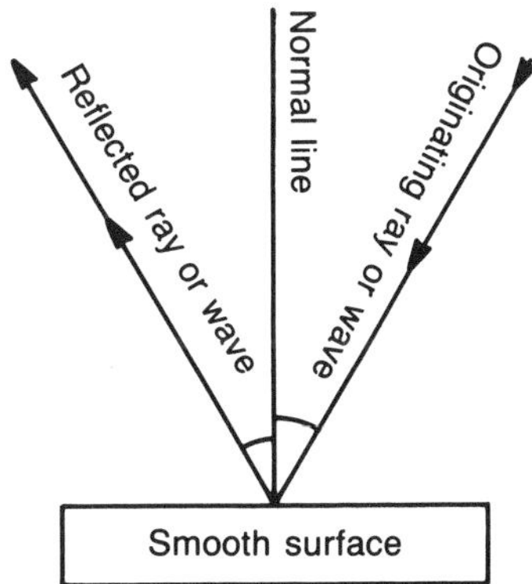


Fig. 2-8. Reflection is basically the return of light (or some other form of electromagnetic radiation) from some surface back to the medium from which it has just traveled.

Table 2-4. Properties of Light.

| Property | Description |
|---|---|
| Reflection | Light returns from any surface which it strikes back to the medium from which it has just left. Its angle of reflection is equal to the angle of incidence. |
| Refraction | Light deflects as its speed changes during passage from one transparent medium to another. The index of refraction is equal to the ratio between the speed of light in a vacuum and the speed of light through the medium. |
| Diffraction | Light bends around the edge of an object or bends as it passes through a very narrow opening. |
| Interference | Light alternately produces bands of lights and shadows as it passes through narrow and parallel openings. |
| Polarization | The wave vibrations of light always take a definite and predictable form. |

noted, change this relationship between angle of incidence and reflection so that we might possibly observe a difference in color between the incidence beam and the reflected beam.

The property of light which is second most noticeable in its tendency to refract. This refers to the way in which light rays bend depending upon the medium through which they travel and the speed at which they may be traveling, for light moves at different speeds through different media. The way in which light moves through a prism is one example of this. Another example would be the macabre illusion a cameraman gets when his lens is aimed at someone standing in a body of water and half the lens is below the surface; the resulting image is of someone with his head somewhat removed from his body.

The third most immediately noticeable fact about light is its tendency to diffract; that is, light rays begin to spread whenever they speed by the edges of some object, despite the fact that light, when acted upon by some outside force, will travel in straight lines.

Less observable, and generally only so in a laboratory environment, are the interference and polarization facts about light. Interference refers to that phenomenon by which rays of light will alternately produce bands of lights and shadows whenever the rays pass through narrow and parallel openings. Polarization refers to the fact that light has its electric vector oriented in such a way that it is highly predictable. This does not mean, however, that there is no such thing as non-polarized light, but it is much easier to produce polarized light than unpolarized light. Polarization is actually well-beyond human detection except with the aid of technology—technology as simple, sometimes, as a pair of sunglasses. The rainbow is linearly polarized, for its electric vector lies in a plain. A pair of sunglasses strike out part of the rainbow's spectrum and thereby facilitate vision or protect the eyes.[17]

Understanding the reflective properties of light has been necessary to the development of lenses as well as to the advancement of observational astronomy, but these reflective properties have never brought physics closer to understanding the very nature of light, nor has the refractive properties of light. However, investigations of the phenomena of diffraction, interference and polarization have made researchers re-evaluate their position on whether light was a particle or a wave. Rather than taking one stand or the other, they took the middle ground, but not out of indecision. The results of all the research since the latter part of the nineteenth century have indicated that light is related in many ways to particle phenomena and also to wave phenomena, as Einstein explained. It is clearly evident that light has both wave-like and particle-like properties, and when you stop to consider the very different items that can be a means for generating light or for detecting it, it becomes quite evident that light is indeed a strange occurrence.

As for the many different items that may generate light, you see natural sources like the Sun and contrived sources from the genius of man. All light generation, however, is dependent upon heat, so there must necessarily be a direct relationship between temperature and light. Light-generating objects are basically bodies containing large populations of electrons in highly activated states. Light can be intensified, then, by increasing the body temperature of a source. Nature does this as a matter of course in the Sun and other stars. What this process or system that we personify as Nature does in stellar objects, is to affect heavy interaction between atomic nuclei. This interaction occurs because the particles involved have been stimulated into colliding, and these collisions often result in the production of new particles different in makeup from the source pair. This process is called nuclear reaction and it is Nature's way of maintaining the intense temperatures we find on stars. Before men and women ever understood the most elementary basics of heat and light or their relationship, they stumbled on a way to harness fire. Then, curiosity, necessity, creativity and courage led to ideas like torches, candles, lamps and, the electric bulb. In pre-history, they produced fire with the most elementary understanding of chemistry; later, thousands of years later, they produced fire with what they understood about electric currents. In all cases, men and women have always needed heat to produce light; they heated elements like wood or fiber or wire.

Actually, light-emissions can be produced by a number of processes and Nature—or, rather, whatever intelligence lies behind the apparent design in nature—has a number of processes for making objects luminous. Some of these means are chemiluminescence, photoluminescence, ionoluminescence, and electron luminescence. Chemiluminescence usually comes about as the result of low-temperature chemical reactions as is demonstrated by the light coming from a lightening bug whose system produces a bioluminescence as a result of the slow oxidation of its organic structure. Photoluminescence, an electromagnetic luminescence, occurs as the result of shorter wavelength radiation, like x or gamma rays, striking some substance in such a way that it produces light. Ionoluminescence is produced by high energy particles, like protons. Electron luminescence is produced by the motions of electrons through a gas cloud.

## ELEMENTARY PARTICLES

"But," you may say at this point, "though I know a little bit more about this strange phenomenon called light, I still do not know what is meant by a particle and what is meant by a wave. Yet, light is some combination of both of these."

In the next part of this book, you will read about how astronomers analyze waves to determine what information they represent about distant objects from which they may have originated, or through which they may have passed on their way to Earth. So, it is certainly important to understand something about wave theory. And because light has many of the characteristics of a particle, particle theory must also be understood. Let's start with particles, because they are a bit easier to grasp.

In its most basic definition, a particle is a fundamental unit of matter. Theoretically, there can be nothing more basic, though every particle need not have the same mass. A slightly more complex, and classical, definition is that which defines a particle as some physical body having measurable mass but negligible extension. Each and every particle that makes up an object has both the properties of inertia and gravitation. Now, just about everything we experience, seems to possess these two properties under some conditions, even some gigantic objects like stars and planets. Because of its negligible extension, a particle cannot be forced into rotational acceleration, and therefore its motion is accepted as being purely transitional.[18] (Transitional motion means the uniform motion of a body in a straight line.)

It would seem, on first thought, that what physicists describe as a particle is very much what philosophers refer to as a monad, or as the soul—a very basic unit of existence. And in a way this assumption is somewhat correct, except that in the

Table 2-5. Some Elementary Particles.

| Family Name & Particle | Mass (electron = 1) | Mass (MeV) |
|---|---|---|
| Photon | | |
| Photon | 0 | 0 |
| Electron | | |
| Electron | 1 | 0.511 |
| Electron neutrino | 0 | 0 |
| Muon | | |
| Muon | 206.678 | 105.654 |
| Muon neutrino | — | 0 |
| Meson | | |
| Pion | | |
| Positive | 273.18 | 139.59 |
| Neutral | 264.2 | 135.0 |
| Kaon | | |
| Positive | 966.6 | 493.9 |
| Neutral | 974.2 | 497.8 |
| Baryons | | |
| Nucleon | | |
| Proton | 1,836.12 | 938.213 |
| Neutron | 1,838.65 | 939.507 |
| Lambda | 2,182.8 | 1,115.36 |
| Sigma | | |
| Positive | 2,327.7 | 1,189.4 |
| Neutral | 2,332.0 | 1,191.5 |
| Negative | 2,340.5 | 1,195.6 |
| Xi | | |
| Neutral | 2,556.0 | 1,311.0 |
| Negative | 2,580.0 | 1,318.0 |

case of a particle, we are dealing with something which is a minute subdivision of matter, whereas with ideas like monad or soul, we are dealing with things which are not expected to be a part of the physical world.

Now, when a physicist uses the word "particle," it is a synonym for the term "elementary particle." Thus, the particles to which we are now referring are what may be considered the very basic constituents of matter. Specifically, they are subatomic units of matter and energy (such as photons, gluons, leptons, and quarks), which are characterized by a certain mass, charge, and spin. These elementary particles, in their individual states, or by their interacting with other basic constituents, bring about the matter with which we exist and coexist. So, particles are subatomic existences which form the basis for everything we know in the physical world, from microbes to stars.

Table 2-5 is a list of the members of the five known families of particles. Some of these elementary particles have been verified in laboratory experiments, others are merely theoretical existences. Every particle has an antiparticle. Antiparticles, few of which have actually been discovered, are mirror images of their corresponding particle; however, their electrical charges are opposite those of the particle. When particles and antiparticles come into contact, they transform into pure energy, perhaps into gamma rays. Theorists assume that the universe has equivalent amounts of particles and antiparticles. As of yet, there is no proof of this theory; but the search goes on. The only antimatter in the solar system appears to result from certain high-energy collisions. The solar system, otherwise, is a planetary system of only matter. But somewhere there must be objects or systems with enough antimatter to balance all the matter in existence. Just why our solar system should be so devoid of antimatter is another of the great cosmological mysteries. One theory has it that if the universe started with a Big Bang, there was probably an inequivalent clumping of matter and antimatter, so that there may now actually be in existence entire galaxies of antimatter just as there are entire galaxies of matter. If these electrically opposite galaxies ever come close enough to each other, they will entirely annihilate each other—or so the theory goes.

The atomic structure of any element consists of a central core, called the nucleus, which contains positively charged particles and uncharged particles. The positively charged particles are called *protons* and the uncharged particles are called *neutrons*. Orbiting about this nucleus are negatively charged particles; these are called *electrons*.

The protons and the neutrons are collectively called *nucleons*. The mass of the atom and the total mass of the nucleons is always equal. Some atoms are referred to as *heavy atoms* because they contain a greater number of neutrons than protons. Just how stable an atom is depends upon the degree in which the neutrons exceed protons; and the ratio here is also important in determining the radioactive properties of an atom.

Radioactivity refers to the spontaneous decay of the atomic nucleus and *this disintegration is often accompanied by electromagnetic radiation*. But, actually, there are three types of radiation that result from a decaying atom. These are *alpha particle* radiation, *beta particle* radiation, and *gamma* radiation. Gamma radiation has already been discussed. Alpha particle radiation is produced from the protons and neutrons in the nucleus, and beta particle radiation is produced from a high speed electron or by a positron.

The negatively-charged particles in orbit about the nucleus (the electrons) are ordinarily in sufficient number to balance the positive charge generated by the nucleus, thus rendering the atom neutral. Change the number of electrons and the atom becomes ionized, which means that the atom has been converted into an *ion*—or, in other words, is now a member of a group of atoms distinguished from all other atoms because they now carry either positive or negative charges. Discovered in 1897, the electron is now realized to be the cause of chemical properties in matter.

*Neutrinos* (not to be confused with neutrons) are associated not only with the electron family of particles but also with the *leptons*. Never detected until 1956, neutrinos nevertheless were believed to exist for more than two decades before their discovery. They are stable particles which are created or destroyed only by the decay of other particles which involves the weak nuclear force, one of the fundamental (but short-range) forces in nature and the force which is capable of influencing the characteristics or motion of a body. *Muons* are all unstable members of the particle population and they generally make up part of the cosmic radiation that may be found near the Earth's surface. The name *lepton*, which defines the family to which muons belong, are a group of particles that are not as massive as either mesons or baryons and are not involved in strong interactions.

The *meson* family of particles are strongly interacting but unstable nuclear particles that have a mass between that of a proton and electron. The *pions* are short-lived and primarily responsible for the nuclear force. The *kaons* are highly massive particles produced in high-energy particle collisions.

The *baryon* family of particles may be distinguished by their rotational characteristics. They all have exactly the same spin. The *nucleon* is either a proton or a neutron, the *lambda* is an uncharged particle that decays into a nucleon and pion, the *sigma* may exist in neutral or charged (+ -) states, and the *xi* is either negative or neutral with a mass more than two-and-a-half times that of an electron in either its negative or neutral state.

The photon is in a family all its own. It has a single positive electric charge and there is at least one proton in each and every atomic nucleus. The photon is a basic measure of radiant energy.

The physical force or energy with which the known elementary particles, containing both mass and charge, interact produces the electromagnetic phenomena discussed earlier (radio, infrared, and other types of electromagnetic radiation). Electromagnetic radiations are generally unaffected by any specific atomic or molecular structure when the medium is immaterial, but when the medium is material, electromagnetic effects are greatly altered.

To sum all of this up, when reference is made to the particle characteristics of light, or electromagnetic radiation in general, reference is being made to its tendency to behave much like what physicists call elementary particles, and specifically a particle called the photon, or *light quantum*.

This particle theory is derived from two types of physical phenomena touched upon earlier in this chapter and in the last. The first is the photoelectric and the other is blackbody radiation. The photoelectric effect refers to the type of emissions that come about as the result of electrons being emitted from the surface of certain substances, particularly metals, whenever light strikes their surface. These emissions are actually the result of electromagnetic radiation being absorbed in the ultraviolet to infrared ranges of the electromagnetic spectrum.[18] Blackbody radiation, discussed briefly on previous pages, is thermal radiation, and refers to the fact that solids and liquids will radiate electromagnetic energy. The actual amount of power that may be radiated will be brought about by the current temperature.[19] Unless light can be explained as being energy transferred by some type of particle, namely the photon, then there is currently no way in which physicists can account for either the photoelectric effect or blackbody radiation.

## THE PHOTOELECTRIC EFFECT

Discovery of the photoelectric effect is sometimes attributed to Heinrich Hertz but the effect was actually discovered in 1899 by the German physicist, Phillip Eduard Leonard (1862-1947), a professor at the University of Kiel (1898-1907) at the time of his discovery. He was also a professor at the University of Heidelberg (1896-1898, 1907-1931) and a 1905 Nobel Prize winner for his research on the properties of cathode ray tubes. When Leonard was conducting his experiments, he was working with something called monochromatic light. This type of light was particularly important to Leonard's discovery because this type of radiation has a relatively small range of wavelengths. What he was surprised to find was that, regardless of the degree in which he illuminated the surface of a metal, the electrons which bounced out all carried the same amount of energy. If he used a light source which had a more intense frequency and, therefore, shorter wavelength, the electrons being emitted from the surface would still produce equal amounts of energy, but this energy was in excess of what was produced by more intense illumination at lower frequencies. Leonard could not come up with a sufficient explanation for this phenomenon and the scientific community had to wait for Einstein to express his ideas.

Einstein prepared a mathematical model which, generally, explained that light beams are actually particle streams and when these particles interact with an atom, the atom gives up an electron carrying the same degree of energy as the particle which made contact with the atom—less whatever energy was expended when the atom gave up the electron. Bright lights mean a greater population of bombarding particles and, therefore, a greater number of electrons being given up. These electrons, however, will always have the same amount of energy as the particles in a beam of light, so no increase in the energy of the electron can be noticed until the particle streams are generated at a higher frequency. These particle streams of which Einstein spoke are actually streams of photons, the photon being the fundamental particle of light (and of all electromagnetic radiation, actually.)

But, at the same time that light has these very definite particle characteristics, light remains a type of physical phenomenon that will change quantatively and qualitatively as it moves in time and through space. This leaves it to be regarded as a wave phenomenon, also.

## ELECTROMAGNETIC WAVES—AN INTRODUCTION

Waves are a completely different type of phenomenon than particles. Waves are basically disturbances that can pass through a medium without having any permanent effect on that medium. By "disturbance" is meant the way in which an environment can be altered or affected in any possible way. Waves are generally described by their amplitude—or motion, that amplitude described in terms of the maximum magnitude of the wave's disturbance.[19]

An easy example of how wave motion can move through a medium and yet leave it fundamentally unchanged is that of a water wave. Actually, in a way, light and water have the same behavioral characteristics; perhaps that is why electronic engineers like to explain electric theory by comparing electric current with the flow of water. Water waves are used here only symbolically. Electromagnetic waves will propergate through many different types of media; but they require no medium for their propergation.

If you are shining a light at a wall, the wall itself is illuminated with reflected light but the space behind the wall is in shadows. If you have a raft or boat in water and the wind and current are hitting one side of the boat, the boat bounces the water waves back toward the direction from which they came while, at the same time, protecting the area on its opposite side from being disturbed (Fig. 2-9). Despite the fact that there has definitely been a disturbance in the water—easily recognizable by the cresting waves or the rocking boat—the water, as you must surely realize, will not be permanently displaced in any way. Likewise, neither will the air which served as a medium by which the light traveled from its source to the wall.

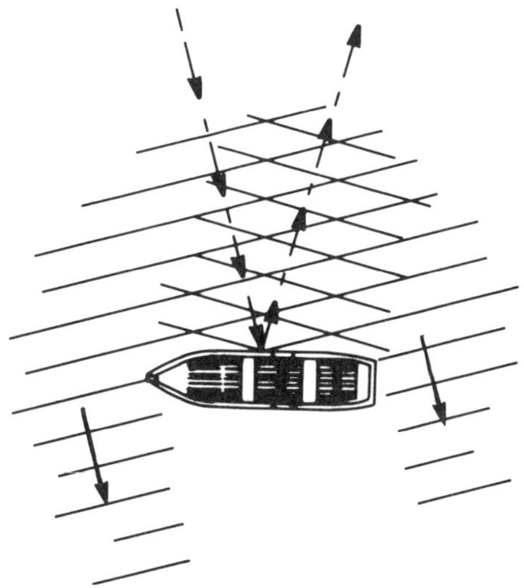

Fig. 2-9. We might relate wave phenomenon to the action of waves on water. Here we see the waves meeting an obstacle and then being reflected back. But, of course, water waves are physical, while light waves are pure energy.

An electromagnetic wave—or any type of wave, for that matter—has both its wavelength and frequency related to a third quantity called the *velocity of propergation*. The relationship between these three quantities is such that the velocity of propergation is equal to the result derived by multiplying frequency times wavelength. In all wave motions, the frequency is a constant except for the Doppler effect, which is fundamentally a change in frequency due to motions at the point of origin or at the point of observation. This is to say, the frequency remains constant unless either the place of origin of the signal or the observer herself move toward or away from each other. Sound and light are both wave phenomena, so comparative examples can be made to illustrate this point.

If you are walking down a street and a speeding car approaches with its horn blowing steadily, there is a steady increase in the pitch as the car speeds toward you; but once the car has begun to pass you, the horn still blowing, the pitch decreases as distance increases. A similar occurrence prevails in the case of light waves, although the occurrence is not exactly the same as it would be for sound waves. In the case of light, first of all, it is necessary to have reference points at astronomical distances, so we must reach to the stars for an example instead of using the headlights of an oncoming, then passing, vehicle. If a star is moving towards Earth, observers would expect that light signals would be received in a shorter period of time than if the star was traveling away from Earth. This means, in effect, that the light rays being received from the star would be more violet than red. If another star happens to be moving away from Earth, then the rays of light from this second would represent the opposite end of the spectrum, and the light rays would be more red than violet. The Doppler effect is a bit different in the case of light than it is in the case of sound, although the fundamental principle is the same. In the case of light waves, it makes little difference whether the source is moving toward the observer or the observer is moving toward the source; for sound waves, however, this does make a difference, for now if the source moves toward the observer, the wavelength will be somewhat diminished. Also, there is no effect on the frequency of light waves if the medium in use is in motion; but there is an effect on the sound waves. Finally, if the observer moves at right angles to the direction of the incoming signals, there will be a change in the frequency of the light waves, although there will be no change in the frequency of the sound waves.[20]

Electromagnetic waves from all regions of the spectrum are expected to act just as do visible light waves. Energy transformation occurs through a process called absorption, in which the radiant wave is changed from one form to another as it moves through various types of material. The thickness of the absorbing material determines the amount of energy that is allowed to pass through it. Just as with visual light waves, the shape of a surface affects the reflection of the wave, the length of the wave governs its refractive tendencies, spherical spreading will cause diffraction, and wavelength (frequency or energy) causes diffraction and scattering.

There are numerous detectors for sensing electromagnetic waves: the bolometer, a resistance thermometer generally used for infrared waves; the crystal, a transparent quartz used in the radio spectrum; photographic emulsion, photoelectric devices, and photoconductor instruments, which are all used in other parts of the electromagnetic spectrum.

The nearest object to us which emits in most areas of the spectrum is the Sun. It is a source of a great deal of electromagnetic radiation, as well as a source of electrons, protons and heavier particles of the atomic nucleus. Most of the radiation from the Sun, however, comes at the visual wavelengths. The shower of electron and proton particles that it sends to Earth have an iconization effect on the upper part of Earth's atmosphere.

## READING RADIATION

In astronomy, the data carried on electromagnetic waves from the infrared through the gamma ray wavelengths is analyzed by spectroscopic techniques. Radiation at the shorter wavelengths depends on space age technology except in those cases where detection equipment is carried in high-altitude balloons; but radiation from the infrared through the near-ultraviolet lengths of the spectrum can still be detected from land-based observatories.

Spectroscopy is a branch of physics which concerns itself with the investigation, and sometimes the production, of spectra. It is concerned with the way in which matter and electromagnetic radiation interact. In optical spectroscopy, a telescope fitted with a spectroheliograph is used to collect radiation. All this spectroheliograph actually does is to disperse the collected radiation into a spectrum which astronomers can have photographed or simply graphed out.

The term "spectrum" which has been used throughout this chapter refers to any graphic display of particle or acoustic radiation. In the case of spectroscopic studies of starlight, or cosmic light in general, it primarily refers to the radiant intensity of the light quantum (photon) as some measurable physical function, like wavelength, or some other quantity. In investigations of electromagnetic radiation, then, the term "spectrum" refers to a series of radiant fields according to wavelength and, therefore, also according to frequency.

There are a number of different classifications for spectra, depending upon the source, the method of generation, or even field parameters. There are absorption, band, continuous line, and emission spectra. An *absorption spectrum* is indicative of radiant energy (or sound waves, but radiant energy is the topic of concern here) that has been intercepted by matter. There is always a certain amount of energy lost during the absorption process but always a corresponding gain in energy by the atomic or molecular structures in the medium. Just how much radiation can be absorbed by any given medium, such as air or water, will depend upon the atomic structure of the medium as well as the density of the matter absorbing the radiation. A *band spectrum* is a well-defined range of electromagnetic wavelengths or frequencies that are generally generated by molecular gases or chemical compounds. A *continu-*

*ous spectrum* occurs when there is a visual light source. The *emission spectrum* may be the result of temperature increases at the source, electron and ion interactions, or the absorption of the light particle (photon).

Spectroscopy has enabled astronomers to understand a great deal more about the Sun, and, as the Sun is a star, to understand more about stars in general. It has also allowed astronomers to gain a great deal of knowledge about the more distant members of our solar system. It has been especially important because of the data it has made available about the Milky Way Galaxy and neighboring and distant star systems. Radiation from the very distant worlds and world systems is received at varying wavelengths from different regions of the electromagnetic spectrum, but astronomers have learned how to interpret the meaning behind the types of radiation being generated. For instance, there is a direct correlation between the surface temperatures of celestial objects and the color they give off, so color readings of spectral data help determine the temperatures of stars. And combining what they know about the Sun with the physical theories that they have developed about stellar objects, astrophysicists can also decipher the chemical makeup of a star from its atmospheric spectrum. Additionally, color magnitude curves reveal the evolutionary stage that a star may be in, and therefore helps in the determination of a star's mass and age.

In short, astronomers and physicists have been ingenious in deciphering the information which radiates throughout all of space. What they have succeeded in doing is to break an important cosmic code, a code that contains the means for unlocking many of the secrets of the universe. They have clearly illustrated through their fantastic ingenuity that *nature is a great communicator, and a great recordkeeper*.

## PAST, PRESENT, AND FUTURE ALL AT ONCE

The most striking thing about the fact that astronomers and physicists have come to understand that radiation carries information and that this information can be deciphered, is not so much that it adds to our greater marvel of the accomplishments of science, for we already take it for granted that the mind of man probably has little limitation. The most striking thing is that this understanding adds to our image of nature as historian and teacher. The ancients well recognized nature as a great teacher and Aristotle and Plato, in particular, believed that the motions of all heavenly bodies—indeed, the entire heavens—were created so that man might learn from its harmony and purpose how to make his own life joyous and useful.

Before nature may be appreciated as this great historian, or recordkeeper, it is first necessary to consider something which is called the *conservation of energy*. This principle states that while energy can be transformed from one state to another, it cannot actually be created or destroyed. This text has primarily been concerned with one type of energy electromagnetic energy—but energy can exist in a number of other forms, including mechanical and chemical forms. The important point about the concept of conservation of energy is that it means the energy in the universe is always quantatively the same, though it may change from kinetic to positional (potential) form. Energy is not the only quantity that remains unchanged in the universe, so does linear and angular momentum and electric charge. However, for the point to be made about nature as great recordkeeper, only the infinite nature of energy need be used as example.

Regardless of what happens in the universe, no matter what stars are born or what stars die or what chaos seems to reign, the amount of kinetic plus potential energy always remains the same. Nature seems to be reluctant to give up anything. It allows change, redefines objects, but it never allows anything to pass into oblivion. Everything happens again, though in some cases its form greatly changes or it is no more than a small part of something greater. Yes, this does seem to guarantee people a type of immortality, but whether or not that will be a personal immortality (in that we remain conscious of "self") is a subject for philosophers and theologians to investigate and debate; it lies outside the realm of science.

Now, if you can understand that nature never actually loses anything but, rather, keeps everything in some form or another—or in so many pieces scattered here, there and everywhere, it is a bit easier to understand how nature can function as this great recordkeeper of universal events. You and I have a great number of ways in which we keep our own personal or business records: we write, we record, we film, we memorize. Well, nature has its own special means of keeping records, not the least of which is fossilization and various chemical clocks. Take, for instance, the history of Earth. This history is contained not only in the writings of men of the arts and sciences but it is also in Earth itself—and in the radiant energy that the planet emits into outer space for some future generation, or some members of a distant planet, to one day decipher, just as astronomers now read the history of the Sun and stars by the radiations they detect from these sources.

As science develops new ideas, the technology to put these ideas to work, it learns to read in greater detail the natural hieroglyphics carved in rocks of fossils on or inside the planet. There are messages about the past all around us. Earth is constantly telling us all about itself. Filed away in trees, in boulders, in soil, in its interior, in its geography, in its chemistry, is Earth's complete history. Science is just beginning to learn how to find the files in which these records are contained, and to decode them. Through radiometric dating, petrology, chemistry, paleonthology and historical geology—just to name a few of the disciplines—men and women are learning that nature is indeed a recordkeeper of the highest order. And through studies of electromagnetic radiation coming from celestial objects, astronomers and physicists are further emphasizing that nature is keeping detailed records of everything that is happening in the universe. Astronomers, as you have now learned, can read the colors of a star and take it as an autobiographical report on the star's age, density and chemical makeup. All around us are records of the past. Science must, however, learn to detect these messages and decode them.

Nature uses a tree to record past events, as trees, by their structure, not only give us their time of birth but also the precipitation and temperature for each and every year of its life.[21] Scientists also suspect that trees can give us records of past forest fires and early frosts. Trees clearly have a way of remembering things and scientists are taking great pains to find out how questions can be asked of them, questions about almost everything that has occurred in the past, from historical dates to earthquakes and volcanic eruptions.[22] Everywhere that it can, nature stores its records, and in such a way that there is enough redundancy to assure no record is completely lost. But for whom does it keep all this data? For you and I and our children? And why does it keep such records? Is it so that we can know the future—and plan for it—by what has happened in the past? And how much more will scientists learn as they begin to read the Earth and sky?

Now, just as nature keeps its storehouses of information at many different locations, sometimes recording different events for the same time period in places across the world from each other, so does it play the role of reporter and teacher. Its role as reporter we can clearly understand, for all about us we can sense the things that are happening, sense the present and the near future, as when storm clouds gather to report on the coming weather. But its role as teacher is not always so evident to us. Yet, in every message it has about the past, it is teaching us what lies in the future. By comparing how things are today with the way they must have been in the past, scientists can observe the patterns and predict the future. Nature seems to want to prepare us for the future.

And nature goes way out of its way to give us hints of the future as it tells us of the past. Just as the humble tree with no mind of its own can help us to understand the reasons for past climatic disturbances and help us to understand the phenomena which caused them, or stars can tell us where from and why they have come, every other thing in nature must contain passages about Earth and cosmic history.

In Greek mythology, there were oracles who warned what the future might bring if things continued the way in which they were going. The lucky man or woman to receive the prophecy now knew in which direction to go to hide from the impending threat or to participate in an event; he or she also knew what to do differently, or have others do differently, to change what the future was about to bring. All around us are oracles, giving us the same warnings. But these oracles are not voices in a well but growth rings in trees, messages in rock, and rays from the Sun.

There is no telling how intricate nature's cross-filing system is. But one example of how interdisciplinary it is may be found in a very interesting article which appeared in the August 1986 edition of *Scientific American*.[23] The article was written by the principal research geologist in the minerals-exploration department of an Australian company who is also an adjunct professor of planetary sciences at the University of Arizona. In the article, the author explains how we might be able to learn about the climatic history of the Earth by laminated sandstone and siltstone formations in a South Australian creek bed. The rocks contain laminated patterns in various degrees of thickness possibly representing cyclical variations of temperatures on Earth some 700 million years or so ago. The rocks are known as the Elatina formation and according to author George E. Williams, will help scientists in evolutionary studies of the Earth's atmosphere, in studies of the solar cycle and in the "prediction of future solar activity."

That Nature brings us the past, the present and the future all at once is also emphasized in a book by Charles Pelligrino called *Time Gate: Hurtling Back Through History*.[24] In it, the author takes the reader on a journey to the beginning of time and does so by treating the reader to the tremendous data that has been deciphered from the physics of Earth, detected from radiation signals in the sky, discovered in the physicist's lab, or theorized by scientists from many disciplines. The entire book is a very good example of nature as historian and teacher.

## SUMMARY

There is little that can be known about the earliest investigations into the phenomenon of light. The ancients were more concerned with the way in which light was generated, or the way in which vision occurred, rather than in the very basic nature of light.

The Pythagoreans believed that the eye itself held the power of vision. Empedocles taught that the eye generates its own light onto an object and then when the light is reflected back from the object, the eye is able to sense the physical characteristics of the object.

While Euclid continued the ideas of the Pythagoreans, Plato took a slightly different tack. He explained that the eye generates particle streams to an object; when the particle stream meets a light stream coming to the object from elsewhere, then vision occurs.

Democritus of Abdera, however, could not accept that light rays were actually generated by the eye and argued that they were emitted, instead, by luminous objects. Aristotle, on the other hand, theorized that light must be something like a gas and it spreads through the air, lighting it until it gradually loses its energy.

While ancient theories about light were elementary or very wrong, the ancients did design and employ optical lenses.

But until the eleventh century, the ideas of the Pythagoreans continued to dominate the science of optics. Then, in that century, Alhazen proved to his colleagues that vision was possible only when light is generated toward the eye. Roger Bacon developed Alhazen's ideas even further and made considerable contributions to the science of optics. Bacon conducted some highly sophisticated experiments with magnifying lenses, but he never advanced to constructing a telescope.

Studies in the physics of light began to become more and more important as a result of the questions amateurs and professionals in the field of astronomy began asking. Their questions were not about light but about stars; however, they knew their questions about the heavens could only be answered if they knew more about light, for what else were the heavens but systems of lights. If they could know its dimension, its speed, its constituents, they could understand more about the sky. Perhaps the first scientists responsible for joining the science of astronomy with the science of light was Johannes Kepler. A book he published in the seventeenth century launched the science of optics to new heights of sophistication and inspired the invention of the telescope.

With the coming of the telescope, astronomy, in turn, was also carried to new heights of sophistication. Who actually discovered the telescope is still some subject for debate. It could have been a manufacturer of spectacles, Zacharias Jansen, and his son, Hans, but it was probably Hans Lippershey, a one-time apprentice to the Lippersheys. Credit for the first astronomical telescope, however, definitely belongs to Galileo. Once he pointed his first telescope to the sky, he began to make one new and startling discovery after another. He verified old theories that the Moon is indeed a world with its own mountains and valleys and he was the first to find the four satellites of Jupiter, still known as the Galilean satellites.

Galileo was also the first scientist on record to try and determine the speed of light. Eagerly, he tried to time the distance light would travel from one hilltop to another but these experiments were destined to failure because any distance on Earth would be too short to measure anything as fast as light. The first scientist to actually conduct a successful experiment to determine whether or not light does indeed travel at a finite speed was Olaus Romer, although his results were not very accurate. Coming closer to determining the true speed of light, however, was James Bradley; he calculated the speed at 188,000 miles per second.

The man who did finally give to us the speed of light was Armond Fizeau; and the man who first successfully measured its speed in a magnetic medium was Edwin William Morley. Morley, along with Albert Michelson and D.C. Miller , was also responsible for disproving the ages old theory of a mysterious medium called the *ether*. Other research by these three men gave new impetus to the wave theory of light.

Two noticeable characteristics of light always intrigues researchers. These are diffraction, which is the tendency of any type of radiation to bend around the edges of some object, and refraction, which is the way in which light is deflected when it goes from one medium to another, as in those cases when it races through air and into water or glass. Scientists were eager to find the fundamental laws which governed these phenomena and, finally, in the seventeenth century, the task was accomplished. The Dutch mathematician, Willebrod Snell stated the law of refraction in his observation that the size of the angle of incidence is the ratio of the refracting medium's index of refraction to the first medium's index. And it was the Jesuit priest, Francesco Grimaldi who discovered the law of diffraction. The phenomenon of double refraction, however, in which certain crystals can cause one single beam to become two distinct beams, was discovered by Erasmus Bartholen.

The first significant theory of light also appeared in the seventeenth century. This was the pressure theory of Rene Descartes. Descartes believed that light had the same basic properties as pressure. Descartes also found the sine law of refraction, although the law was named after Willebrod Snellius, who developed it independently and a bit earlier.

Until Isaac Newton's ideas began to take hold, Descartes' pressure theory remained at the forefront. Newton believed that light was made up of elementary particles. Rays of light were not "homogeneal," he explained, and the different rays of light caused color. Newton argued that what Descartes perceived as a steady pressure of light was actually a particle stream moving at exceptional speed.

Newton's ideas, however, wound up taking a close second to Huygens' wave theory. But when Huygens passed away, Newton's particle theory began to excite new, but short-lived, interest. Before long, physicists were back to the wave theory—then back again to the particle theory. And for the two hundred years after the beginning of the seventeenth century, physicists remained unable to make up their minds, alternating constantly between Newton's and Huygen's theories. By the beginning of the nineteenth century, the tide of evidence so supported the wave theory that Newton's arguments were almost completely shelved forever. The knight of science who rescued the particle theory was Maxwell Planck.

Still, a great deal of confusion about light was evident in the scientific community. One special question continued to plague physicists more than any other, and this was related to the medium through which light moved. Whatever this medium was, theorists decided, it had to be some type of elastic solid. This was the only way to explain that light could travel through a vacuum. This elastic solid was called the ether. There was absolutely no evidence for its existence, but physicists were certain that it must exist—certain, that is, until the Michelson-Morley experiments, which disproved the entire idea of the ether.

As it turns out, light has the characteristics of both a wave and a particle. This was partly confirmed by the research of Rudolf Hertz, whose study of radio waves showed that this type of radiation behaved very much like light. But Hertz also found that the amount of electron energy produced by directing ultraviolet light on a metallic surface would be dependent upon the frequency, not the intensity, of the radiation. This gave light the characteristic of a particle. And this dilemma was further compounded by the work of William Conrad Röntgen, whose work with X-rays, which just like visible light are a form of electromagnetic energy, showed that the absorption and emission characteristics of X-rays are corpuscular in nature. And as x-rays go, so must visible light radiation.

By the seventeenth century, it was becoming evident that light acted as both a particle and a wave. Still, the scientific community was reluctant to accept the particle idea; further argument was necessary. These arguments came as the result of Max Planck, who discovered that stars emit and absorb energy in infinitely small "bundles" called quanta. This means that light has to be corpuscular in nature, at least in part. Planck's work was confirmed by Albert Einstein, who wrote that physicists had little choice but to accept light as some special form of physical phenomenon that had the characteristics of both a particle and a wave. And since Einstein's arguments, light has come to be accepted as some form of hybrid that has at least the qualities of a wave and a particle. The basic unit of light is called the *photon*, or light quantum.

Research and theories about light and color that have had any importance may be traced as far back as the first century A.D., and to the Roman philosopher, Seneca. He noticed that there was a great similarity between the colors of the rainbow and light passing through a glass cut with three corners and a sharp edge. This observation brought him to conclude that color was intrinsic to light, different colors occurring when light passed into or through certain objects. More contemporary ideas on the subject of color began with experiments by Isaac Newton, who noted that some "rays are disposed" to exhibit one color and other rays another color. He felt he had enough evidence to prove that refrangibility and color were closely tied together. Up until Newton's experiments and observations, the science world simply accepted color to be some natural necessity way beyond the understanding of men and women. Newton's ideas, however, were never fully appreciated until the nineteenth century.

Thomas Young revived the interest in Newton's work. Young conducted a number of experiments which showed that any number of "new" colors could be brought about by mixing red, violet, and green lights in different proportions. He thus concluded that the human eye is equipped with three very distinctive means of color perception. Young's theories were expanded by James Clerk Maxwell. What was particularly important about Maxwell's experiments, and those experiments which they influenced, was the idea that a beam of light could be spread out for further analyses. This striking idea, which had its roots in Newton's experiments, led to the development of spectroscopy, the science that investigates the effect of electromagnetic waves.

Light is, finally, a form of electromagnetic radiation, as well as an unusual type of elementary particle or "package" of particles. What is electromagnetic radiation? It is radiation transmitted as electromagnetic waves, waves containing both electric and magnetic fields. The energy produced by an electromagnetic wave is actually the result of the acceleration of a charged particle.

The forms of electromagnetic energy, besides that of visible light, which are especially important in astronomy are: gamma ($\gamma$-ray) radiation, X-rays, ultraviolet radiation, infrared radiation and microwave and radar radiation. By analyzing the electromagnetic radiation coming from celestial objects, astronomers are able to tell a great deal about the source.

Knowing that light is both a particle and a wave—or, rather, something that has both the characteristics of a particle and a wave—is of little value if someone does not understand just what is meant by a particle, or just what is meant by a wave. A particle, as we accept the term in this text, is a fundamental unit of matter, so when the physicist uses the word "particle," he uses it as a synonym for the term "elementary particle." There are many families of particles: electrons, leptons, mesons, baryons and photons. This last is the light particle. There are probably many more families of particles still to be discovered. Basically, if light is not a particle, then it is impossible for physicists to explain the photoelectric effect and blackbody radiation. The "photoelectric effect" refers to the way in which electrons are emitted by various objects whenever these objects are illuminated. The term "blackbody radiation" refers to the way in which solids and liquids radiate electromagnetic energy; it is none other than thermal radiation.

A wave is a very different physical phenomenon than a particle; a wave is mainly a physical disturbance that can move through any given medium and not leave that medium permanently displaced in any way. It does not need any medium for its propagation. This explains why light can shine through things and leave objects unaffected, for light is a type of wave, an electromagnetic wave. How can light have the characteristics of two such very different types of phenomena as a wave and a particle? Well, the answer to this question is still to be found. Physicists do not know exactly what light is, but they are sufficiently sure of the characteristics of waves and particles to understand what may be expected of light and how it may be produced.

# Chapter 3
# The Development of Astrochemistry and Astrophysics

ASTROPHYSICS IS THE ACADEMIC MARRIAGE OF astronomy and physics. It deals with cosmic energy, the messages contained in these energy emissions, and the physical laws which govern the universe. The term came into wide use in scientific writings about 1890 but the very beginnings of astrophysics may be traced back more than one hundred years. Generally, historians trace the birth of astrophysics to the discovery of astronomical spectroscopy by Josef Fraunhofer (1787-1826) or else to the founding of photometry by Bouguer when he published his *The Gradation of Light* in 1760. Still others credit the beginning of astrophysics as a formal field of study to the merging of spectroscopic, photometric, and photographic techniques for the purpose of studying celestial objects.[1]

But before this time, Newton had already been conducting studies of how a spectrum was created and how colors were formed; and as he was using direct sunlight in his experiments, one might point to these investigations as the first merging of astronomy and physics. However, even before Newton, Kepler was diverting from his mathematical investigations to consider the nature of light; he no doubt understood that a study of the stars and planets was basically a study of light, and to learn the how and why of the planets and the stars, scientists would have to investigate thoroughly the causes and characteristics of light.

## SOLAR SPECTROSCOPY

The great breakthrough in astrophysics, however, was accomplished by Gustav Kirchhoff (1824-1897) when he succeeded in decoding the Fraunhofer lines. The last chapter introduced Kirchhoff's work and discoveries with Robert Bunsen (1811-1891), whose name is most often associated with the gas burner named after him.

It is always interesting to learn how one discovery leads to another, and that sometimes the birth or development of an idea is the result of investigations that were on a different track altogether. And so it was with Bunsen's burner and Kirchhoff's investigations.

The particular importance of the Bunsen burner was that it burned without giving off any colors.

**Table 3-I. Astrophysics and Astrochemistry.**

| Field | Techniques/Technology | Subject Areas |
|---|---|---|
| Astrophysics | Observational; utilizes different instruments depending upon areas of electromagnetic spectrum being investigated; technology includes image tube scanners, image photon counting systems, photographic plates, interferometer arrays and moveable dishes and many types of telescopes. | Generally all subject areas of astronomy, from those concerned with planetary and stellar structure to those encompassing cosmological theories. Astrophysics is not concerned, however, with positional and motional studies. |
| Astrochemistry (cosmochemistry) | Observational (meteorites) as well as laboratory; technology includes chemical analyzers and many of the techniques ordinarily employed for laboratory analysis of chemicals. Much of the work depends upon theoretical models and the reading of radiation from space. | Specifically concerned with chemical compounds and elements in the universe, with formation theories, isotropic and radioactive occurrences and nuclear reactions. |

Now, this is a rather unusual phenomenon because a heated object usually generates visual radiation in colors that will change as the heat becomes more or less intense. Thus, the Bunsen burner presented a unique instrument for observing colors that might be associated with the burning of different chemicals, and Kirchhoff was particularly interested in its applications in chemistry. Kirchhoff immediately started analyzing the possible uses of the burner and Bunsen, as curious as any scientist would be to find out the potential of his invention, welcomed Kirchhoff's curiosity.

To observe the colors in the flame in as much detail as possible, Kirchhoff and Bunsen designed a very special system for their experiments. They got together a couple of telescopes and arranged them so that one could be used to collect light from the flame and direct it through a small opening to a prism, and the other could be used to observe the resulting spectrum from the prism.[2] Their little system of instruments was, of course, no more than a slight variation of Fraunhofer's spectroscope.

Kirchhoff and Bunsen eagerly observed the spectral details before them but were somewhat amazed to find that the gas burner produced a separate pattern of bright lines rather than a continuous pattern. And, as they were soon to learn, every different chemical substance that they burned produced very different spectral patterns, but each chemical always produced the same pattern. (It should now be easy to understand how spectroscopic studies of astronomical objects can reveal much about their chemistry.)

Years before, Joseph von Fraunhofer had been conducting solar studies and came to discover the very dark lines that appear in the solar spectrum (now called *Fraunhofer lines*). Fraunhofer's observations revealed that there were close to 800 of these dark lines, although now astronomers have catalogued more than 25 thousand. In the yellow region of the solar spectrum, Fraunhofer noted a pair of very prominent lines that he called "D-lines" and noted that similar very marked lines appeared in the same region of the visible spectrum of sodium. (D-lines actually occur at wavelengths of 589.6 nm and 589.0 nm.) But Fraunhofer was not able to demonstrate a direct relationship between the D-lines in the different spectra. However, he was convinced that the lines and bands that were produced were due to the nature of sunlight and definitely not from the properties of light or some optical illusion.

Inspired by his preliminary investigations of the Sun, Fraunhofer went on to conduct spectroscopic studies of the Moon, Venus, Mars, and the stars. He found that their spectra had the same characteristic lines as the solar spectrum. Additionally, he was somewhat amazed to find that solar and starlight had the same refractibility in every color. This proved to him that the light coming from the Sun and the stars was most likely produced in the same way.[3]

Kirchhoff, remembering Fraunhofer's work, hoped that Bunsen's burner could help him determine if there was indeed any relationship between the solar and sodium lines. He decided that the most direct way in which he could find the answer would be to produce a solar spectrum and observe the resulting lines when a flame colored by sodium vapor was put in front of the slit in his spectroscopic system. When he did as planned, he immediately observed that the dark D-lines now became bright lines.

Next, he tested to see how the intensity of the solar spectrum would affect the sodium lines; he did this by shining as much sunlight as he could through the sodium flame and into the slit. The result was an exceptionally clear display of the D-lines. "I then exchanged the sunlight for Drummond's oxyhydrogen limelight," he later explained, "which would produce a spectrum containing no dark lines, as would any incandescent solid or liquid body." When he directed this light through a flame already colored by common salt, where the sodium lines once existed, there were no dark lines. Kirchhoff explained that the reason for this was that the sodium flame absorbed rays which had the same degree of refrangibility as the sodium rays, but to all other rays it was absolutely transparent.

Kirchhoff continued the same experiments over and over again, but each time colored the Bunsen flame with a different vapor. His experiments led to the fomularization of the following laws related to spectroscopic analyses: There is a specific and distinct spectrum for each chemical species; and every element absorbs the kind of radiation it emits, an event called the reversal of the lines.

By studying the relationship between the spectral characteristics of elements used in his laboratory experiments and the spectral characteristics of the solar atmosphere, Kirchhoff was able to begin developing a chemical profile of the Sun. The Sun, he showed, was much like the Earth in chemical constitution, and consisted not only of sodium but also of calcium, copper, zinc, barium, nickel and other familiar chemical elements. Additionally, he was able to demonstrate that the solar photosphere was of much greater temperature than atmospheric zones.

## DEVELOPMENTS IN ASTROCHEMISTRY

Kirchhoff's experiments were actually responsible for launching another discipline, which might be viewed as either a subdivision of astronomy or cosmology or geophysics. This is the field of astrochemistry, which is concerned with both the organic and inorganic chemical makeup of not only stellar objects but interstellar space as well. Astrochemists depend primarily on spectroscopic investigations as a means for conducting their studies. However, they have sometimes been privileged to have actual samples to analyze. In the past, this was because they had meteoritic samples (and they still do); now, thanks to the space age, they've had access to lunar samples, also.

Astrochemistry is still in its infancy, depending a great deal upon theoretical models to tie together the relatively scant physical evidences with radiation studies. This means that at this stage of its development, astrochemistry is very dependent upon the optical and invisible astronomies discussed in the second part of this book. Basic chemical studies of stellar and interstellar objects of lower temperatures is carried on at the visible, infrared and ultraviolet wavelengths of the electromagnetic spectrum. The more temperature intensive regions of space and of specific objects, however, is carried out in the X-ray and gamma ray regions.

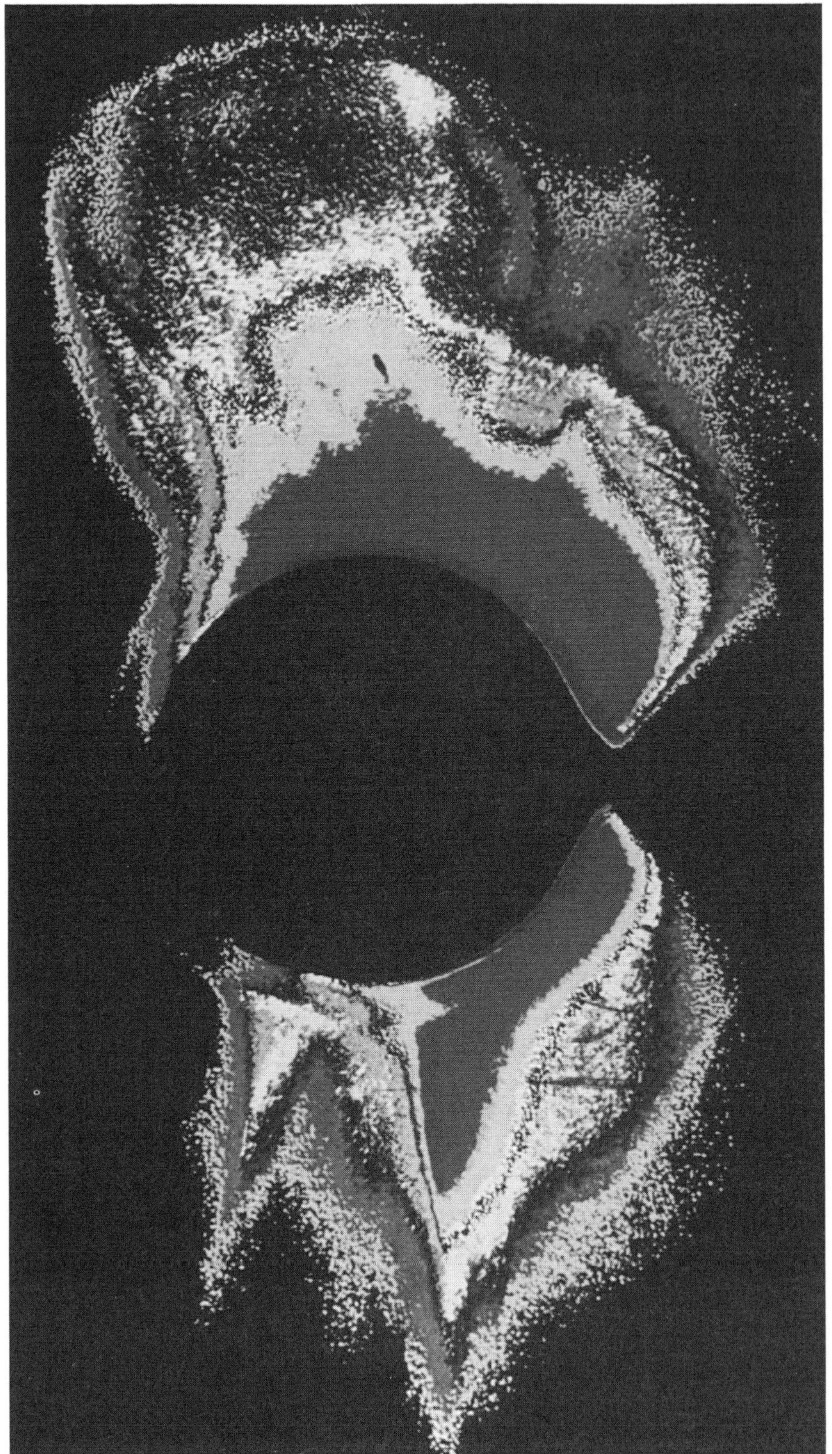

Fig. 3-1. The Sun's corona, color-coded to distinguish brightness levels. This photo was taken by a coronagraph on Skylab. A coronagraph artificially eclipses the solar surface through a darkened occulating disc at the center of an imaging lens' focal plane.

Since the invention of the spectroscope, a great deal of data has been compiled about the chemistry of the solar system and a great deal has been amassed about other galactic systems. The Sun, astronomers have learned, consists mainly of hydrogen, then helium, then quantities of oxygen, nitrogen, carbon, magnesium, silicon, iron, sodium, sulfur, aluminum and calcium. These same elements and more are found throughout the universe. Based on studies of solar radiation, geological studies here on Earth, and analyses of meteorites, astrochemists have developed standard cosmic abundance tables. These must be at most guidelines, for there are just too many stars to study, and there is just not enough knowledge about solar physics to allow the compilation of final tables. Astrochemists generally prefer to base their cosmic estimates on the data they are able to obtain from meteorites. As they see it, meteorites have greater ties with the beginnings of the solar system and, therefore, with the fundamental nature of the universe. If cosmological theories are correct, Earth has emerged from the great pri meval cloud that formed the solar system during a rather static cosmic period and so, therefore, is not as indicative, as meteorites are, of the way things were in the early phases of solar or galactic evolution.

The chemical composition of the Earth is not necessarily indicative of the chemical composition of the other eight major planets. But as one of the four "terrestrial" planets, its chemical properties are at least similar to that of Mercury, Venus, and Mars. These planets show evidences of an abundance of magnesium, silicon, iron, aluminum, calcium, and chromium. In the case of the Earth, however, iron is usually in an oxidic rather than a metallic form, which represents a different circumstance than that for meteoric rocks in which the iron is in oxidic as well as metallic forms. The core of Earth is probably some combination of oxygen, silicon and sulfur. Earth, by the way, is one of the three of the nine major planets that has an atmosphere. The other two planets are Venus and Mercury.

Mercury is the closest planet to the Sun and probably the second smallest. (Astronomers are not sure of the size of Pluto.) It is about 4,880 kilometers in diameter, making it about one-third the size of the Earth. The surface temperature of Mercury is about 450 degrees Celsius when it is orbiting within 46 million kilometers of the Sun. Its nighttime temperatures, however, can drop to below 200 degrees Celsius. Because the mean density of the planet is very high, astronomers guess that it has an abnormally large core, probably consisting mainly of metallic iron.

The next planet from the Sun is Venus. This is an intensely hot planet, even hotter than Mercury though Mercury orbits much closer to the Sun. The reason for the extraordinarily high temperatures is in the planet's cloud covering. The clouds are so dense that they create a greenhouse effect in which infrared rays are trapped to heat the surface below as well as the atmosphere. Light from the Sun making it through the dense cloud covering is absorbed and then radiated back at extended infrared wavelengths. The mainly carbon dioxide atmosphere

**Table 3-2. Planetary Physics.**

| Planet | Radius (At Equator) | Volume (Earth = 1) | Mass (Earth = 1) | Density (Earth = 1) | Probable Aver. Temperature |
|---|---|---|---|---|---|
| Mercury | 2425 km | 0.055 | 0.053 | 0.99 | + 600 F |
| Venus | 6051 km | 0.087 | 0.815 | 0.95 | + 900-1000 F |
| Earth | 6378 km | 1.00 | 1.00 | 1.00 | + 70 |
| Mars | 3395 km | 0.150 | 0.107 | 0.71 | − 10 F |
| Jupiter | 71600 km | 1317.00 | 317.89 | 0.23 | − 240 F |
| Saturn | 60000 km | 762.00 | 95.15 | 0.11 | − 300 F |
| Uranus | 25900 km | 67.00 | 14.54 | 0.23 | − 340 F |
| Neptune | 24700 km | 58.00 | 17.23 | 0.31 | − 370 |
| Pluto | 3000 km? | .12? | .002 | 0.25? | ? |

Table 3-3. Planetary Motions.

| Planet | Sidereal Revolution (In Years) | Mean Velocity (km/sec) | Orbital Eccentricity | Surface Gravity (Earth = 1) | Inclination |
|---|---|---|---|---|---|
| Mercury | 0.241 | 47.89 | 0.20564 | 0.37 | 7 00 17 |
| Venus | 0.615 | 35.03 | 0.00676 | 0.88 | 3 23 40 |
| Earth | 1.000 | 29.79 | 0.01676 | 1.00 | 0 00 00 |
| Mars | 1.881 | 24.13 | 0.93348 | 0.38 | 1 50 59 |
| Jupiter | 11.862 | 13.06 | 0.48061 | 2.64 | 1 18 19 |
| Saturn | 29.458 | 9.64 | 0.05082 | 1.15 | 2 29 08 |
| Uranus | 84.014 | 6.81 | 0.04755 | 1.15 | 0 46 27 |
| Neptune | 164.790 | 5.43 | 0.00661 | 1.12 | 1 46 15 |
| Pluto | 247.700 | 4.74 | 0.25336 | 0.04 | 17 07 56 |

(The last column refers to the inclination of the orbit to the ecliptic. Also, note the following comparative statistics: the planets closest to the Sun are moving with the greatest velocity, one year on Earth is equal to 247 years on Pluto, and that Jupiter has the strongest surface gravity.)

does not let the rebounding infrared rays escape into space, but instead forces the rays back to the surface. Atmospheric re-emissions meet the surface emissions, causing temperatures to increase drastically. Space probes have revealed that this very thick atmosphere consists mainly of carbon dioxide (96%). They have also detected small amounts of nitrogen and minute traces of sulfur dioxide, water vapor, carbon monoxide, argon, helium, neon, hydrogen chloride, and hydrogen fluoride.

The next planet from the Sun is Earth, and beyond Earth is Mars. Mars is a relatively small planet,

Table 3-4. Gases Detected in Abundance in Planetary Atmospheres.

| Gas | Planets |
|---|---|
| Acetylene | Jupiter |
| Argon | Mercury, Earth, Mars |
| Carbon Dioxide | Venus, Earth, Mars |
| Carbon Monoxide | Venus, Mars, Jupiter |
| Ethane | Jupiter, Saturn |
| Helium | Mercury, Venus, Earth, Saturn |
| Hydrogen | Mercury, Jupiter, Saturn, Uranus, Neptune |
| Hydrogen Chloride | Venus |
| Hydrogen Fluoride | Venus |
| Krypton | Earth |
| Methane | Jupiter, Saturn, Uranus, Neptune, Pluto |
| Neon | Mercury, Earth |
| Nitrogen | Venus |
| Nitrous Oxide | Earth |
| Oxygen | Earth, Mars |
| Ozone | Earth, Mars |
| Phosphine | Jupiter, Saturn |
| Water | Venus, Earth, Mars |
| Xenon | Earth |

though big in reputation thanks to sci-fi writers who often make it the base for alien attacks against Earth or the environment in which their great tales take place. Mars actually has an equatorial diameter of only 6,787 kilometers. It is the last of the inner planets and is separated from Jupiter by the asteroid belt, that system of minor planets that separates the terrestrial planets from the gaseous and giant planets in the solar system. The atmosphere on Mars is extremely thin and consists mainly of carbon dioxide and relatively small amounts of nitrogen, argon, oxygen, and water vapor. Atmospheric clouds are mainly carbon dioxide and ice.

The outer planetary group, which consists of Jupiter, Saturn, Uranus, Neptune, and Pluto, shows a markedly different chemical structure than the inner planets. These planets orbit the Sun from beyond the asteroid belt, some 740 million to 7½ billion kilometers from the Sun. Except for Pluto, these are giant planets four times to twelve times the size of Earth.

The largest planets in the solar system are Jupiter and Saturn. Jupiter orbits between Mars and Saturn, has a mass over 300 times that of Earth, and, with its equatorial diameter of almost 143,000 kilometers, is the largest planet. Saturn has an equatorial diameter of roughly 120,000 kilometers and is about 95 times greater in mass than Earth.

Because it radiates heat at about twice the rate it absorbs it from the Sun, Jupiter is suspected of having a reservoir of thermal heat beneath its yellow gaseous surface; its core is probably made of iron and silicates and is probably tens of thousands of degrees in temperature. Saturn's surface area is similar to Jupiter's; and like Jupiter, it is a cold planet with a very powerful magnetic field. Both Jupiter and Saturn have chemical compositions very similar to that of the Sun. In some ways, they appear to be stars that have become planets (if this is possible) or else planets on their way to becoming stars (again, if such a thing is possible). Both Jupiter and Saturn are predominately composed of the elements hydrogen and helium. In the case of Jupiter, the extreme interior pressures keep the hydrogen in a metallic state, though closer to the surface one discovers a sea of liquid oxygen. The core of the planet is probably iron silicate in molten form. As for Saturn, its interior is expected to be much like Jupiter's, so theoretical models picture it as a planet with a very dense core surrounded by metallic hydrogen. However, though it is smaller than Jupiter, Saturn appears to send off energy emissions comparable to those of the larger planet. This makes astronomers surmise that there is a relatively large helium content in the hydrogen field surrounding the core. This would not only account for the unexpected energy emissions but also for the relatively small amount of helium content in the atmosphere.

The other great planets, Uranus and Neptune, are markedly different not only from the terrestrial planets but from Jupiter and Saturn as well. Much of the data gathered about Uranus is relatively new, the result of the unmanned explorations of NASA's Voyager 2, now on its way to Neptune. Uranus is the seventh of the major planets (in terms of distance from the Sun) and is roughly 52,000 kilometers in diameter. Neptune is almost exactly the same size. It is believed that the chemical compositions of these planets are identical but this cannot be verified until Voyager 2 makes it safely to and around Neptune. Both planets have solid rock cores packaged in ice perhaps 8,000 kilometers thick, and hydrogen running just about as thick. These are cold planets, just like Jupiter and Saturn. Like Saturn, Neptune is radiating more energy than it is receiving from the Sun, indications that it, too, has an internal energy source; but in the case of Neptune, astronomers suspect that the energy is the result of potential-to-kinetic conversion of gravitational energy. Optical data exhibits strong evidence of methane as well as hydrogen absorption. Theoreticians propose that the planet is also made up of significant amounts of helium, neon, ammonia, hydrogen, sulfide, and other compounds.

Pluto is another matter altogether. As far as anyone can tell, this is the most distant planet in the solar system, and one that was only discovered a few decades back. C.W. Tombaugh was hard at work in February, 1930 conducting an exhaustive photographic search for a planet whose existence had been predicted by Lowell and others. Finally, he came across a humble speck that quite notice-

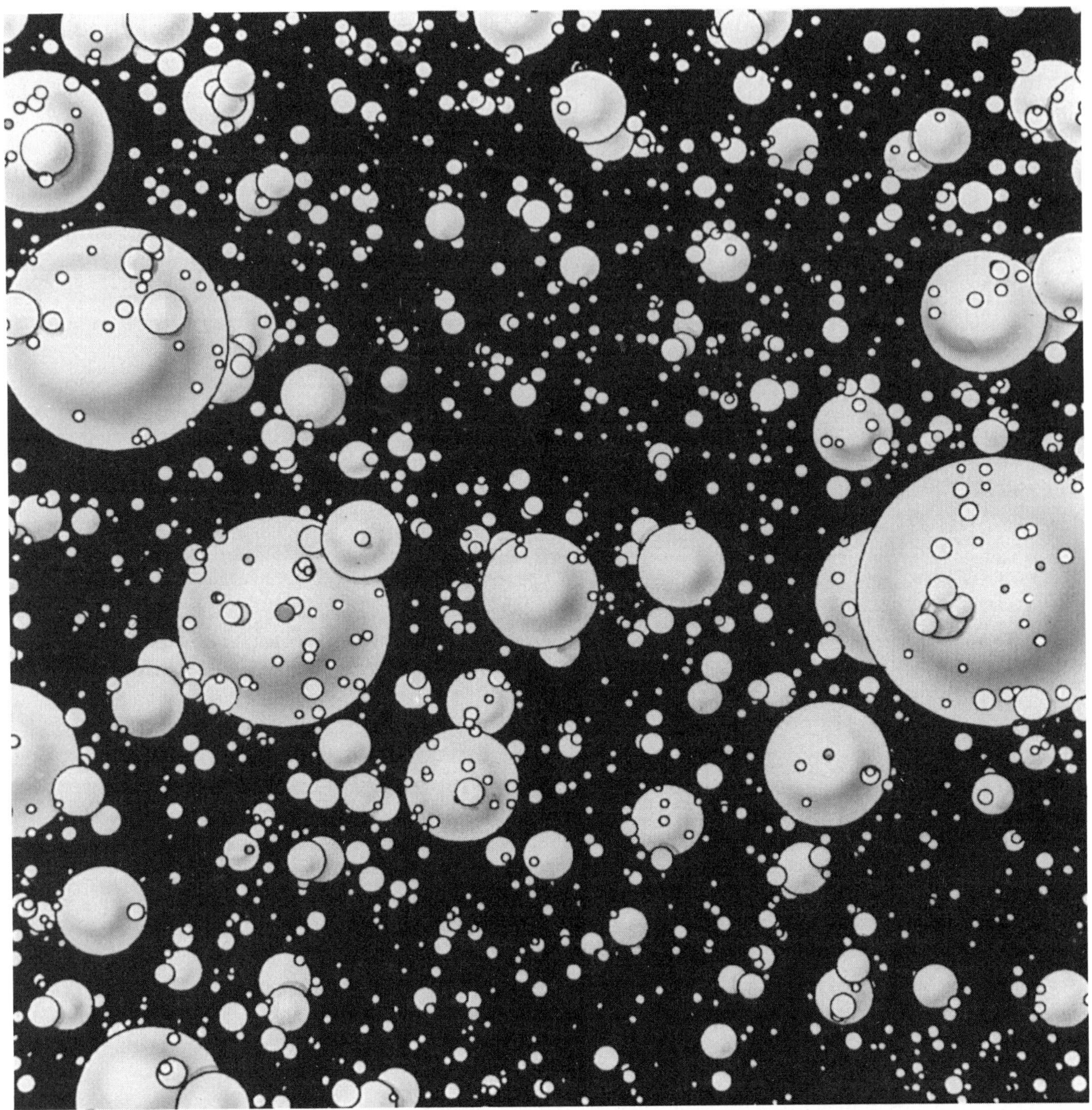

Fig. 3-2. This is a computerized representation of the number and size of the particles populating a 3 × 3 meter area of Saturn's outermost ring. The particles actually range in size from about two centimeters to 70 centimeters. There are other particles not represented here which are much larger than 70 centimeters. The number of particles decreases rapidly as size increases (Courtesy of Jet Propulsion Laboratory).

ably had changed positions in a short period of time. Lowell, Pickering and others had long suspected the presence of Pluto because of the marked perturbations in Neptune's motion, but discovering the planet and learning about it are two completely different challenges. Pluto is so very far away that there is little that can be found out about it. The documentation about it has been piling up, but most of the information is theoretical. Some astronomers theorize that Pluto may have been a satellite of Neptune

Fig. 3-3. The planet Neptune was believed to exist long before it was eventually observed. The man who calculated its position while it was still an unknown planet was John Couch Adams (1819-92) (Courtesy of Library of Congress).

that finally escaped gravitational slavery to the planet. This theory rests primarily on the six day rotational period and the high orbital inclination and eccentricity. As far as can be determined, the planet only has a diameter of some 4,000 kilometers and it appears to be made up mostly of ice. Spectroscopic studies reveal the presence of methane in solid form and many astronomers guess that it is also present in gaseous form. It will be a long time before astronomers can add much to the very little currently known about tiny Pluto. It is much too far for even Voyager 2 to reach, and certainly much too small for effective study from Earth stations or spacecraft.

The system of dwarf worlds making up the asteroid belt between Mars and Jupiter is actually a system of planets, for this is exactly what asteroids are: planets. However, this is only because they orbit the Sun. They are hardly big enough to be habitable even if they could develop habitable environments. Many of them are a mere thousand kilometers in diameter. While it is true that they orbit the Sun just like Earth and the other major planets, their orbital eccentricities are much more pronounced than those of the major planets. There may be 100 thousand or more of these asteroids (or *planetoids*) but at this time there is no way to be certain about exactly how many of them are in existence. Some theories explain that at one time these asteroids were more than simply orbiting rocks. They may have been at least the size of Pluto or one of the satellites in cosmic ages past, but their highly eccentric orbital patterns brought them to collide with each other and they were fractured, broken or blown apart. Some of their sections were hurled out of the solar system, some collided with the other planets or moons, and some of their sections became these things we call asteroids. To determine the surface composition of these asteroids, astronomers have made spectral studies of the sunlight they reflect. Generally, they are classified in two compositional groups, one consisting of carbonaceous materials and the other olvine silicates mixed with metallic iron (which may be the solid form left after a comet blows itself up).

**Table 3-5. Some of the Major Asteroids.**

| Asteroid (Minor Planet) | Diameter |
|---|---|
| Ceres | 1025 km |
| Vesta | 555 km |
| Pallas | 538 km |
| Hygiea | 450 km |
| Interamnia | 338 km |
| Davida | 335 km |
| Cybele | 311 km |
| Europa | 291 km |
| Patentia | 281 km |
| Euphrosyne | 261 km |

An asteroid is suspected of having created the Everglades when it collided with an ancient sea bed some 38 million years ago, according to Edward J. Petusch of Florida International University. And a report in *Astronomy* magazine indicates that there is mounting evidence to support Petusch's theory.[4] Magnetic readings by the Florida Bureau of Geology detect metallic ore beneath the muddy surface of the Everglades, which, until Petusch's theory, was thought to be carved out by groundwater. The impact is believed to have occurred somewhere in the vicinity of what is now Miami. At first the crater caused by the collision was filled with seawater, but as millions of years passed, the rising coral walls that had been forming since shortly after the catastrophe blocked off the sea and in the natural course of events, fresh water replaced the seawater. It is very possible that many other lakes, swamps, or craters owe their existences to asteroid collisions.

Occasionally confused with meteoroids by the general public, comets are the great solar wanderers. They orbit the Sun much like the major and minor planets but they are probably ice-filled celestial constructions, although current cometary investigations indicate they may possibly have other than icy cores. Some astronomers suggest that comets have their origins in the solar birth cloud. The great nebula that gave birth to the protosun and the protoplanets is believed to have contained excessive amounts of ice. When the protosun began to increase intensely in temperature, the ice that had formed in the inner parts of the cloud was burned away. At the very extremes of the birth cloud, however, hydrogen gas combined with other materials; as a result, methane, ammonia and water ice was allowed

**Table 3-6. Solar System Objects Easily Confused.**

| OBJECT | DESCRIPTION |
|---|---|
| Asteroid | Also called *minor planets* or *planetoids*, these are large rocks that orbit the Sun much like the nine major planets. The largest is about 1000 kilometers in diameter. Their total mass is just about $^{4}/_{10,000}$ths that of Earth. |
| Comet | A relatively small object consisting of a nucleus, coma, and tail that orbits the Sun like minor and major planets but which has a highly eccentric orbit. |
| Meteor | A small dust particle burning itself out as it falls through Earth's atmosphere. |
| Meteorite | Interplanetary objects that crash to Earth, some large enough to cause craters, others small enough to be no more than bouncing pebbles when they reach Earth. |
| Meteoroid | A term encompassing all the interplanetary debris resulting from collisions in meteor streams. |

to form. In this environment, comets could thrive. And so it is believed that perhaps a billion years or more ago, they were in tremendous numbers, forming their own cometary belt at the fringes of the solar system. Spectral studies have lent support to this theory, as they have shown that comets are made up of water, methane, ammonia, and carbon dioxide. Those other wayward objects which are sometimes mistaken for comets—the meteoroids—are celestial rocks that move about the Sun in low inclination direct orbits. There are theories that some meteoroids are actually pieces of asteroids that have broken away from the main body and are now hurtling about the solar system in their own orbits. All that can really be seen of them is the light they produce as they race about the Sun. Meteoroids, meteors, and meteorites are actually all one and the same thing in different circumstances. When meteoroids crash-land on Earth, exploding into oblivion as a result, they are called meteors; if anything is left of them after the landing, they are called meteorites. Although they are essentially rocks, they are very different from the rocks found on Earth. As meteorites, they are generally classified in three ways: *stony* meteorites; *iron* meteorites; and *stony-iron* meteorites. The stony meteorite is somewhat like Earth rock but consists mainly of oxygen, silicon, magnesium, and iron, with oxygen the predominate element. The iron meteorite is 89 percent iron, with nickel making up most of the rest of its composition. The stony-iron meteorite has equivalent amounts of nickel-iron and silicate.

Recently, meteorites have been found that contain diamonds. These are, of course, microscopic diamonds, but this discovery is a rather monumental one in astronomy. It indicates that interstellar dust contains diamonds and also that because of the durability of these microscopic samples, the diamonds may actually contain traces of other materials that will help astronomers understand stellar composition. That meteoritic material sometimes contains diamond particles is nothing new in astronomy, but until now it was believed that these particles were created from the force of collision on Earth. Now, however, astronomers are certain the diamond particles were picked up as the meteorites passed through stellar environments and that these diamonds were manufactured at relatively low pressure conditions.

When these meteors are no more than bright lights streaking across the sky, they are often referred to as "shooting stars." These shooting stars are fairly common occurrences, and millions can be seen in the sky each and every night, although not all by the same observer or from the same position. The distance from which these shooting stars can be observed will depend on their size and brightness; but for most shooting stars, the first-sight zone is roughly 150 kilometers above sea level. Those meteoroids that become meteorites fall to Earth at about 10-35 kilometers per second. At about 100 kilometers from Earth, they begin to disintegrate. If they do not explode or melt, they may be reduced to 10 percent of their original size. The meteorites are slowed by the Earth's atmosphere so by the time they make it to the surface, they no more than bounce to the ground to finally rest a mere meter or two below the soil. Sometimes the meteorites are large and fast enough to do a great deal of damage; many craters are attributed to meteorite impacts.

Right now the origin of meteoroids is a great mystery. Astronomers are hard at work constructing theories and the evidence to support those theories. Meteoroids may be related in some way to asteroids, or even to comets, but for right now we must wait for startling discoveries, or theories that can withstand the onslaught of scientific challenge, before we can know their reason.

Meteoroids, asteroids, comets, planets—we have now looked at the members of the solar system and considered their chemical compositions. This leaves us now to advance further into the Galaxy and to the stars and interstellar space.

Stars tend to have chemical compositions which are related to their evolutionary period. The evolutionary program for any given star is generally determined by the star's mass, if current theories are correct. At different stages in the development of a star, nuclear reactions will greatly alter its basic constituents. The relative amounts of chemicals change as the star increases in age.

The chemical structure of interstellar areas of the Galaxy varies depending upon whether studies are being made of those areas where new stars may be in the process of developing or in regions where solar age or older stars exist. Most of the interstellar gas and dust floating about the Galaxy is clumped in the spiral arms where the new stars are believed to be developing. The intense emissions from these galactic arms usually burn away the dust and the existing gas is converted to heat energy, thus producing ionized hydrogen. Outside the vicinities of the newly emerging stars, the gas is relatively cold, roughly 60°-90° K (Celsius) and a different chemistry ensues. A broad range of interstellar chemistry, then, is detected in galactic surveys. In some regions there will be high percents of calcium, sodium and iron; in others there will be high percents of hydrogen, carbon, nitrogen, and oxygen.

As far as can be detected, most galaxies appear to have chemical makeups quite similar to our Galaxy. However, not all galaxies have the same mix of stars in their populations; many of the very far galaxies contain a large number of very old stars. This is because these very distant galaxies evolved first, and with them their star populations. Different stellar ages mean different chemical readings. Very active galaxies, however, present a specially unique problem to researchers, for a great deal of their radiation is synchrotronic, which means that astrochemists cannot extract chemical information from the emissions.

## DEVELOPMENTS IN ASTROPHYSICS

After Kirchhoff's experiment, interest and experiment in astrophysics began to accelerate. The first major investigations concentrated on the Sun. In the 1860s, the Scottish physicist and natural philosopher, David Brewster (1781-1868), noted for his studies of the polarization of light as well as his invention of the kaleidoscope, observed that the Earth's atmosphere produced very narrow absorption lines in the solar spectrum. These are called telluric lines and are very strong when the source, solar or otherwise, is near the horizon. Brewster first noticed them when he was doing spectral studies of a setting Sun. He was not sure, however, about which atmospheric elements were causing the telluric lines, but studies by French astronomers indicated later that water vapor, oxygen and carbon dioxide were all significant causes.

In 1869, the Swedish physicist, Anders Jöns Ångström (1814-1874) published the results of his solar spectral studies. Ångström was very famous for his studies of light and in particular for his analyses of spectra. He published important works on not only his optical research but also on his studies of spectra of simple gases. His 1869 publication on the solar spectrum listed the wavelengths of 1,000 lines, expressed in terms of $10^{-10}$ meters, a unit of measure that has since become known as the *angstrom unit*.

Throughout the rest of the century, other scientists continued Angström's work. These included Henry Draper, the American physiologist, astronomer and New York University professor, who had built his own observatory at Hastings-on-the-Hudson; Sir William de Wiveleslie Abney (1843-1922), the British chemist and physicist who did a great deal of experimental work in photography, and was the first to produce photographic plates that were sensitive to red and infrared radiation; Marie Alfred Carnu (1841-1902), the French physicist who contributed to studies on the velocity of light, and on acoustics and optics; and Henry Augustus Rowland (1848-1901), the Johns Hopkins professor who developed spectroscopy into an exact science.

By the 1860s, new theories about the Sun were beginning to crowd scientific literature, all built upon the basic observations of Kirchhoff. The astronomical community was now interpreting the Sun to be a giant world of gas with all of its heat being generated from somewhere in its interior, and was extremely interested in pursuing photographic studies of the solar surface. The most impressive of these efforts at photography during the early stages of solar research were those taken by the Mauden Observatory in 1876; these photographs were of the granular structures populating the photosphere.[5]

Solar research still remains at the forefront of astrophysical research. But it is now not simply concerned with the solar surface and atmosphere but with all astronomical phenomena related to the Sun. This means that solar physicists are concerned not only with the star itself but with how the star effects other celestial bodies and general celestial events, including the aurora and magnetic storms we witness here on Earth. The solar wind is also of particular interest to physicists. This wind developed in the solar corona when a flux of protons, electrons and some heavier elements are heated into a force so powerful that it produces shock waves in Earth's magnetic field and influences cometary tails. But, then, the Sun is a source of tremendous energy of all kinds, just like any star of its mass and age. Its primary energy is the product of a series of thermonuclear explosions which occur near or at its core. During these thermonuclear explosions, hydrogen is converted to helium. As the conversion takes place, energy is created but mass is destroyed. It is this process that keeps the Sun glowing and capable of heating our solar system, but these thermonuclear explosions take their toll. As the Sun slowly loses its mass, it is slowly giving up its life. Physicists predict that at the present rate at which the Sun is converting hydrogen into helium, the star may only be around for a trillion years or so, possibly only hundreds of billions of years. You and I need not worry about events so far in advance, unless we are philosophical and enlightened enough to realize that if the human race can last that long, it may not have a chance of lasting longer unless we support astronomical research and space travel. Just as we are here today because of atmospheric developments on Earth billions of years ago, perhaps others will be around billions of years from now because of what we can learn in time to save them. When the Sun finally dies, so will the entire solar system, most of it having already deteriorated to barren rocks or having been burned away when the Sun entered its final evolutionary period. Hopefully by then, men and women will have migrated to another solar system, another galaxy—or, possibly, another universe.

The 1860s saw the scientific community develop special interest not only in solar studies but also in stellar studies. Here is where the work of men like Secchi, Huygens, and others played important roles in the growth of astrophysics. Pietro Angelo Secchi (1818-1878) was the Jesuit priest to whom you were introduced earlier. He was director of the observatory at Rome's Gregorian University and was well respected for his spectroscopic studies as well as for his astronomical photography. Inspired by Kirchhoff's experiments and observations, he began cataloging low dispersion stars and assigning them spectral classifications. His catalog of some 4,000 stars, which he published in 1869, established four spectral classifications: white, yellow, orange-red, and red. He expected that there was some relationship between the spectral classes and the temperatures of stars but he appeared to have no interest in chemical analyses of the lines and bands in spectra.[6]

Huygens (1824-1910), mentioned earlier, pioneered spectroscopic photography and helped influence the combined usage of telescopes, spectroscopes and photographic equipment in astronomy. Huygens, unlike Secchi, was extremely interested in conducting investigations of stellar spectra and, by 1863, he had successfully determined the composition of a few of the brighter stars in the sky.

One thing quickly led to another, and soon spectroscopic analyses was also playing its role in determining the velocity and age of stellar objects. This came about because of the work of the German astronomer, H. Vogel, who compared spectral lines of opposite limbs of the solar disk in order to measure the rotational speed of the Sun.[7]

Since the time of Secchi, Huygens, and Vogel, a great deal of data has been compiled about the physical makeup of the Sun, stars, and planets. The chemical composition of the average star is roughly 70 percent hydrogen, 27 percent helium, 2 percent carbon, nitrogen, oxygen, and neon, and about 1 percent iron and heavier elements. Astronomers now also realize that stars are great, luminous, gaseous spheres held together by their own gravitational interactions, and that they may be anywhere from one-tenth to 50 times the size of the Sun. They are at extremely great distances away from not only Earth but each other, although the illusions we receive from the evening sky makes the stars appear to be rather close to each other. Stellar distances are so great that it is impossible for men and women to travel from one star to another, unless science discovers exceptions to known physical laws that might speed the journeys. Even nearby stars like Alpha Centauri, Barnard's, and Wolf 359 are from four to eight light years away. Right now there exists no means whatsoever for traveling at the speed of light, or anywhere near that speed. Of course, published distances to the stars cannot be very accurate, for there is no direct means for making the measurements; lasers and radar can be used to determine distances within the solar system and space probes can eventually verify those measures, but the distance to objects lying beyond our solar system cannot be determined with any dependable accuracy. But besides being very distant, stars are also very hot. Even if we could find some way to travel to them, we could not get close enough to them for direct observation. Their surface temperatures range from a couple of thousand degrees to possibly 100 thousand degrees and more.

Astrophysicists hope to be able to learn how stars come into being and to fully understand their evolutionary phases. Current theory still supports a model of stellar birth and evolution that begins with the gravitational collapse of very cool and highly dense gas and dust clouds. Gravitational collapse means the contraction of the cloud as a result of its own gravitational interactions. These birthclouds are parsecs in area and consist mainly of molecular hydrogen and, perhaps, cosmic dust in a quantity representing about one percent of the cloud's mass, or they are smaller clouds with diameters of less than seven-tenths of a parsec that seem to prevail in interstellar areas containing large percents of ionized hydrogen. Once the clouds begin to collapse, governing physical laws cause them to separate into smaller cloud formations. This fragmentation process continues until finally clouds are formed which are much too dense and hot for further fragmentation. These extremely dense and extremely hot nebulae represent the beginnings of the protostar stage. These very hot, very dense clouds will continue to collapse but they do not divide. As they collapse over periods of tens of thousands to millions of years (depending upon their mass, density, and temperature), they gradually become hotter and hotter until a young star is formed. While this model of star birth is highly theoretical, studies of electromagnetic radiation coming from "star clouds" seems to support the theory. What next happens to the cloud or the young star cannot be stated so surely, either, and again astrophysicists must work within the realm of pure theory. They expect, however, that a star's future is determined to a great extent by its mass, eventually becoming a white dwarf, neutron star or black hole. (Stellar evolution is further explained in the chapters on X-ray and gamma ray astronomy).

Astrophysics is not limited to solar and stellar

research only. It is also concerned with the solar system, interstellar matter and extragalactic systems or objects.

Studies of the solar system are concerned with interior and atmospheric conditions of the planets as well as the physical makeup of comets, meteors, asteroids and any other objects or galaxies that may be detected in space. Studies of interstellar matter are aimed at unraveling the complex structure of nebulae and other visible and invisible matter that exists between the stars. Studies of systems lying beyond the Galaxy, such as pulsars or quasars, will help astrophysicists and cosmologists learn a great deal more about the evolution of the universe and what may actually exist beyond our Galaxy and the Local Group of galaxies to which it belongs.

Much of the data now being compiled about distant stars and galaxies, and what lies beyond them, is a product of the collection and analyses of electromagnetic radiation given off by celestial objects. Radio, infrared, ultraviolet, X-ray and gamma ray studies are of primary importance to astrophysical research. And microwave detection has played its part in cosmology. In coming chapters, you will see how researchers in visible and invisible regions of the electromagnetic spectrum have joined forces to enhance our understanding of not only the solar system but the entire universe as well. The final answers to all the questions that science poses about the very nature of the universe, its birth, and its evolution, are a long way from us. Perhaps we will never know the answers to the ultimate questions. But astronomers continue to collect and analyze electromagnetic data to add whatever pieces to the puzzle that they can. Optical, radio, infrared, ultraviolet, X-ray, and gamma ray astronomers all have their own types of telescopes and, as you will learn, their own special problems in studying the universe; but they continue, undaunted, led on by their tremendous curiosities and the challenges that await them.

## ELECTROMAGNETIC WAVES—A MORE TECHNICAL DESCRIPTION

The electromagnetic radiation that astrophysicists work so diligently to collect and decode do not actually carry information any more than a pulse in the central processing unit of a computer carries data. An engineer cannot put a pulse under the microscope and find illustrated upon it graphics or alphanumerics relaying a message. But this pulse has been caused by something—by some gating or multiplexing configuration enabled by an event—and if the electronic technician knows the cause of the pulse, he can interpret on those terms. Similarly, if astrophysicists understand the causes of the different types of radiation, they can relate an event or series of circumstances to the presence of that radiation.

All waves, have three properties: length, frequency, and velocity.

The wavelength is the distance covered by a periodic wave motion during one of its oscillating cycles. If you were to be at sea in relatively rough water, you would notice that the water rises to a very noticeable crest, sinks again, and then rises once more. The distance from one crest to the other would be the wavelength; and so it is in the case of electromagnetic waves.

The wave frequency is the number of oscillations that occur over a period of time. As you observe the cresting waves and count the number of waves that pass by you each and every second, you are measuring the velocity of the water waves. In this same way, physicists measure the velocity of electromagnetic waves, except that they express the time in hertz (hz), which is a unit of frequency equal to one cycle per second. The frequency range of electromagnetic radiation is roughly 3,000 hertz at the lower frequencies to possibly $10^{22}$ hertz at the higher frequencies. But hertz is not the only measure used with electromagnetic radiation studies. In measuring waves physicists may also use angstroms, microns ($10^{-6}$) or centimeters.

The wave velocity is the time rate of its linear motion in a given direction. As you observe the sea from the starboard catwalk of your vessel, you will observe that the crest of any wave is traveling in one specific direction. The distance which the crest moves each and every second represents the wave's velocity. To determine the velocity you must know the wave frequency and the wavelength

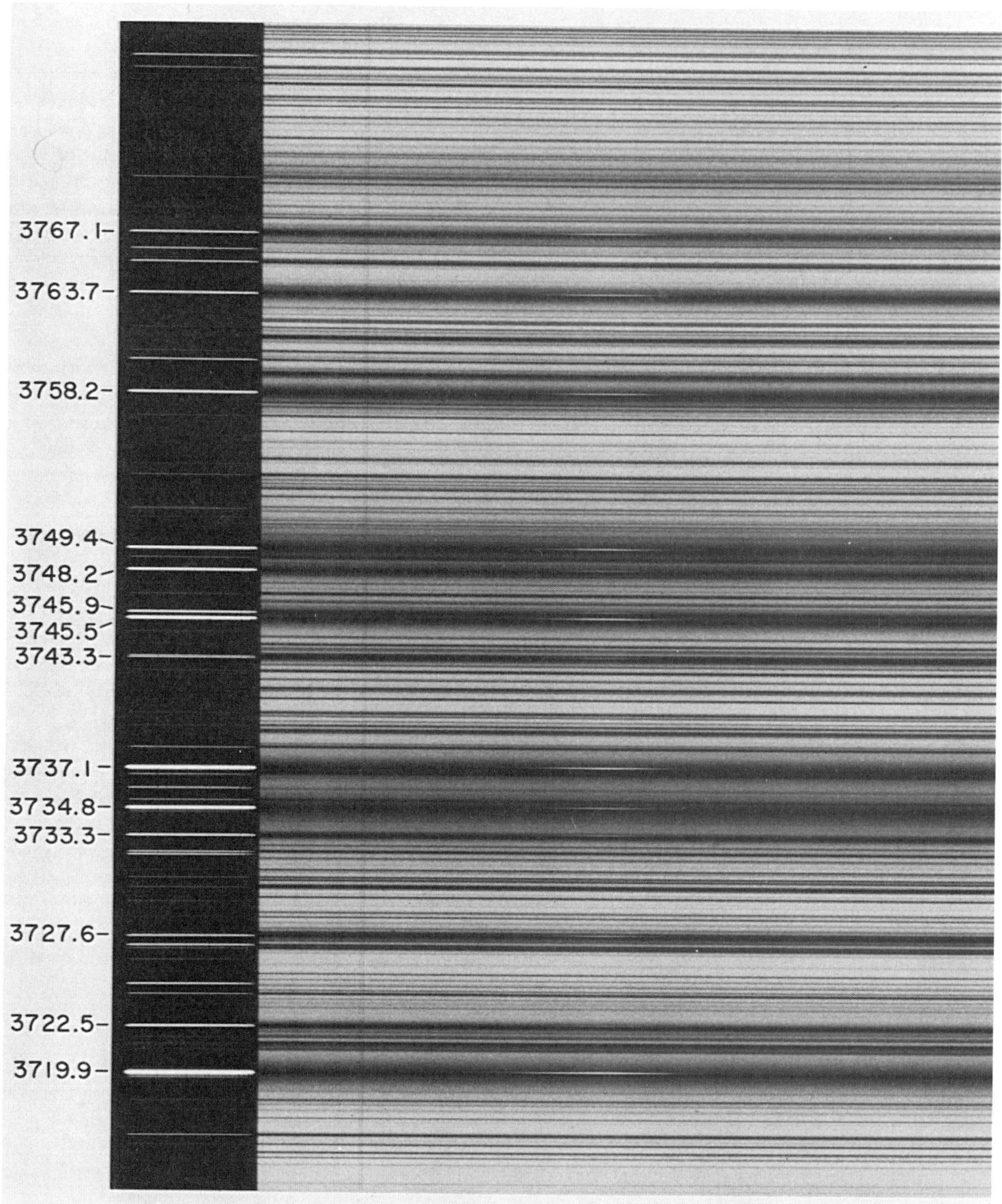


Fig. 3-4. Here is the portion of the near ultraviolet spectrum of a comet, this one being the comet Kohoutek (Courtesy of Lick Observatory).

(velocity = wave frequency × wavelength). Using the same formula, physicists measure the velocity of an electromagnetic wave.

Thus, astrophysicists study the length, frequency, and velocity of an electromagnetic wave in order to decode the information contained on the wave. Waves do not actually carry data, for waves are only an energy force; that is, they are not at all involved in the transportation of matter of any kind but simply represent the transportation of energy. As you watch the ocean, you instinctively know that the waves you see are not carrying the water to the shore, otherwise eventually those waves—and the ones following it—would move the entire ocean elsewhere and regardless of the power of your vessel's engine, you would go the way of the ocean. Ocean and land areas would continually change. It would be a crazy world.

There are big differences, however, between ocean waves and electromagnetic waves. For one, the ocean waves are a visible phenomenon and they need some medium for their propagation, as wind or earthquake. Electromagnetic waves, on the other hand, are invisible, need very sophisticated instruments for detection, and need no medium for propagation. Ocean waves carry and support. Electromagnetic waves are pure energy.

Hopefully, interpretation of the electromagnetic signals being received from celestial objects will help astrophysicists "get to" stars and galaxies long before the physical means to actually do so is available. Once "there", they can begin to unravel the great mysteries confronting anyone who wants to understand the why, how, and when of all things. We have discussed briefly the current theory of stellar birth and evolution, yet there are too many questions which severely challenge these theories, not the least of which is: What are the physical laws that can allow a cloud to contract and fragment into protostar segments?

A lot of the descriptions found in textbooks about astronomy or cosmology are presented so logically and with such great description that we assume quite readily that what we read is fact. But a lot of the ideas we have about the solar system, the stars and the galaxies are just that—ideas. There is so much to learn!

It should be noted that the "voids" that seem to exist between clusters of galaxies may not actually be voids at all. This is underscored by the recent discovery of a gas-rich and extremely dark galaxy in a "void" just beyond the Virgo Cluster. And dark galaxies like this one can very well be one of the answers to the great question about the missing mass in the universe.

# Part 2
# Foregrounds

# Chapter 4
# Visual Light Astronomy

SOMEHOW, THE FANTASTIC INTELLIGENCE BEHIND nature has made a special effort to compensate for man's lack of natural instincts or of intellectual sophistication. This intelligence has provided for man's nurturing, found ways to communicate with him (as explained in Chapter 2), and has developed his environment so that he might not only be physically sustained but mentally exercised.

For thousands or millions of years—depending upon how long you believe men and women have been on Earth—they have been provided with no natural means for detecting radiation from the full range of the electromagnetic spectrum. It is a good thing that this has been so, for just imagine what a mass of confusion life would be if every time someone opened her eyes she had to sort out the wavelengths in which she wanted to operate over any given period. Fortunately, men and women have only been given the ability to detect radiation in the neighborhood of $3 \times 10^{-7}$, although babies may also be able to detect ultraviolet radiation. In other members of the animal kingdom, exceptions have been made and certain creatures can switch from visible light to infrared or ultraviolet light, or else operate solely in another region of the spectrum. For instance, it is believed that bumble bees see in ultraviolet light and that snakes, and possibly other nocturnal creatures, have the ability of infrared detection at night and visual detection at day.

Vision in the $3 \times 10^{-7}$ range of the spectrum may be explained scientifically in terms of evolutionary theories, but the big question is why the potential exists for seeing at these wavelengths instead of others. Nonetheless, from the very beginning of his existence, man has been able to detect at least one kind of radiation and to use this highly developed sense to investigate things well beyond the reach of his other senses.

## THE FIRST GREAT OBSERVATORIES

Today, professional optical astronomy is conducted mainly from great observatories built high above the clouds usually cluttering the lower atmosphere and interfering with the visual radiation from stars, planets and galaxies. Amateur astronomy, however, is conducted from wherever the enthusiast happens to have his telescope and to whatever has aroused his curiosity. The amateur actually looks into his telescope and spies on the sky. The professional rarely looks into his telescope; he is more concerned with computer printouts and spectral data. This is one reason why amateur astronomers remain such an important part of the astronomical community, and continually make such important contributions. They are not limited to specific research goals and can follow anything that catches their eye.

Astronomical observatories are not really new ideas. Even back in the time of the ancient Egyptians and Babylonians, astronomers used the highest rooms in temples or the tops of towers to set up shop and make their observations. Even then astronomers instinctively tried to get away from the man-made lights or lunar reflections that could disturb their view. Probably the very first astronomical observatory was built by Ptolemy Soter (Ptolemy I) around 300 B.C. Ptolemy was one of Alexander the Great's generals who, on the death of Alexander, took control of Egypt and eventually became the first king of the Macedonian empire.

In the middle ages, the Arabs had constructed great observatories in Baghdad (ninth century) and in Damascus (tenth century). The Europeans, however, were somewhat slow to see the need for specially-equipped buildings to which astronomers could retreat in order to measure planetary phenomena and to keep up-to-date on the latest documentation related to their field. The first European observatory of any measurable significance went up in Nuremberg just about twenty years before Columbus made his historic journey. Construction of that observatory was actually the brainchild of the German astronomer Regiomontanus (Johannes Muller) and of his patron, Bernard Walther. Regiomontanus was also a mathematician, and while his first duties were to the Church—he became a bishop in 1475, the year before his death—he found the time to make considerable contributions to the improvement of astronomical instruments and the use of mathematics in astronomy. A little more than 100 years later, Tyco Brahe built an even more sophisticated observatory at his castle on the island of Hyeen, as reported in Chapter 1. These early observatories were little more than great libraries consisting of measuring, calculating, and observing instruments; they were built so that points in their construction could be associated with solar, planetary, or stellar positions on certain dates. Remember that even in Brahe's time, the telescope still had not been invented. This means that it was not until the 1600s that there could be the design and construction of astronomical observatories that were the forerunners of what astronomers have available to them today. It was also during that century that the first efforts were made to construct observatories at public universities and other public institutions.

Fig. 4-1. A quadrant used by the astronomer Hevelius. Hevelius also constructed the largest telescope of his time, one that exceeded the telescope built by Huygens by a length of 27 feet (Courtesy of Yerkes Observatory).

The first observatory established by an educational institution (1637) was the observatory at the University of Leiden, a learning center founded in 1575, and the oldest university in the Netherlands. The city of Leiden itself has played an important role in history. Dating back to the Roman Empire, it was an important industrial center in the middle ages and was the birthplace of von Goyen and Rembrandt. It was typical that the university bearing its name should continue the city's fame by being the first school to have its own observatory.

The next observatory to be built by an educational institution went up on the campus of a Jesuit college in Ingolstadt,[1] but this turned out to be a rather poorly equipped observatory in terms of telescopes. It fell far short of the Leiden Observatory. It had as its primary instrument a 4-foot quadrant, an instrument used for determining angular altitudes.[2] The quadrant was a popular measuring instrument until the eighteenth century when it was replaced by the sextant. The sextant was the invention of Thomas Godfrey and John Hadly. There has always been dispute over who actually deserves credit for its invention but it was decided by the Royal Society that both Godfrey and Hadly had claim as inventors. Godfrey was a bit of a maverick; he was a self-educated astronomer and mathematician and, as a young apprentice, assisted in glazing Philadelphia's Independence Hall.

The same year that the observatory at Ingolstadt was completed, Christian IV of Denmark (1577-1648), prodded on by the astronomer Christian Severin (1562-1647), planned to construct an observatory in Copenhagen. Better known by the name of Longomontanus, a Latin version of the city (Longberg) where he was born, Severin was one of Tyco Brahe's assistants and, at the time of the design and construction of the observatory, was a professor of mathematics at the University of Copenhagen. He joined the university about six years after Tyco's death and remained a professor of the university until his own death. Neither Severin or Christian IV witnessed the completion of the observatory, since it was not finished until 1657, nine years after the king's death and ten years after Severin's death.

In the latter part of the seventeenth century, national observatories were founded in England and France. The English observatory became known as the Royal Observatory and the French building as the Paris Observatory. The Paris observatory was completed in 1671 and the Royal in 1675. With the tremendously accelerating costs of newly designed and larger telescopes, the small, private observatory was now becoming a thing of the past. Yet there were still some impressive private efforts in the nineteenth century. These were the observatories of Sir William Herschel and William Oarson, 3rd Earl of Rosse (1800-1867), the British astronomer and telescope maker (Fig. 4-2). Parson's work with telescopes had a particularly dramatic impact on astronomy because he was the first to find a method to prevent the deterioration of telescope surfaces during the cooling phase that was part of the manufacturing process. He built a reflecting telescope that had a speculum six feet in diameter, qualifying it as the largest telescope ever built at that time.

## THE CHANGING TELESCOPE

The series of developments that led to larger,

Fig. 4-2. Here is the reflector designed and built by the Earl of Rosse. Rosse made numerous observations of nebulae and was the first to detect that some of these nebulae had spiral formations (Courtesy of Yerkes Observatory).

more expensive telescopes, and the subsequent decline in the construction of private observatories like those of Herschel and the Earl of Rosse, probably could be traced back to the ideas and work of Johannes Hevelius (1611-1687) and Christiaan Huygens (1629-1695). Both men were highly impressed with versions of Johannes Kepler's telescope (which produced an inverted view) that produced a true image as the result of using four convex lenses instead of two. They realized the tremendous potential from the new design and argued for greater acceptance as well as further improvement. Kepler had already demonstrated that the more popular lenses having very curved surfaces gave defective images and that the weak objective lenses were actually of greater dependability. What resulted from all these arguments and recommendations was the construction of extremely large telescopes ranging up to twelve feet in length, the increased size necessary because of the way in which a convex lens with little curvature handles light.[3] The light has to be an extraordinary distance from the lens to come into focus. This idea so caught on that telescopes just became larger and larger until Christiaan Huygens, in 1659, finally constructed one that was 123 feet long. This record was soon broken by Hevelius who built a telescope that was 50 yards long. At these great lengths, size was working against the observer rather than working to his advantage. It was terribly difficult to aim these monstrosities and they were easily swayed by the slightest disturbance so, though they gave better images, it was impossible to keep anything in sight for detailed study.

Many changes have occurred in telescope design and manufacture since these early experiments, but the primary purpose of a telescope still remains the same: to collect radiation in order that radiation may be detected as distinctly as possible. Huygens, Hevelius, and the early telescope makes all were concerned with simply visible radiation; now telescopes are designed to collect all types of radiation. Today's telescopes are larger and more efficient than Kepler would have ever dreamed, but fundamentally they are little different from his telescopes.

Optical telescopes, as they are used in astronomy today, come in three basic designs: refracting, reflecting, and catodioptric. Refracting telescopes are equipped only with lenses; reflecting telescopes are equipped only with mirrors; and catodiopric telescopes are equipped with both mirrors and lenses.

In the refracting telescope, the images carried on rays of light must be inverted, and this inversion is accomplished by the objective lens. The light coming into the objective lens is usually as close to being parallel as possible but once it comes through the objective lens, the rays are refracted at angles that will cause the rays at the farthest extremes to eventually intersect on their way to the eyepiece. This would ordinarily result in severe strain on the eye because of the way in which the rays are now directed at it except that the eyepiece can be focused to reorganize the rays in parallel form. This is, of course, an overly simplified example, for here we speak of using a single lens at the objective and a single lens at the eyepiece. In actuality, telescopes are built with a number of lenses making up the objective system and a number of lenses making up the eyepiece system. This compounding is necessary to eliminate the aberrations caused by the refractive tendencies of glass when light passes into and through a medium, and to eliminate the aberrations caused by spherically designed lenses.

Most refracting telescopes are small and are generally used for terrestrial viewing; so today these types of telescopes are used in cameras, surveying instruments and in various types of military equipment, to name just a few applications. Galileo used refracting telescopes in his astronomical studies, as do astronomers today. Galileo's version was, of course, very basic and also utilized a very small objective lens. The refracting telescopes used in astronomy today utilize very large objective lenses, though for maximum viewing efficiency a 3½ foot (diameter) objective lens is about the maximum size to which manufacturers can go and still maintain the level of quality demanded by astronomers. Astronomers always prefer refracting telescopes when they have to greatly magnify objects. The lenses used in refractors must necessarily be very thin, otherwise they absorb most of the light they should be reflecting. If they're too thin, or they sag and result in intolerable distortions.

Reflecting telescopes, where mirrors are used in place of lenses, have certain advantages over the refracting telescope. This is because mirrors do not necessitate the same restrictions in size required by lenses. The reflecting telescopes designed and manufactured today actually utilize numerous mirrors mounted in such a way that the light rays can be managed into focus. Despite the great advantage of reflecting telescopes,they were never really appreciated by astronomers until the late nineteenth century, because of the failure of early attempts to use the telescopes effectively.

One example of a highly publicized failure was the 1868 construction of a 48-inch telescope in Australia. The project had the professional support of the Royal Society, which served in a consulting capacity to advise the colony of Victoria on the particular type of telescope that should be constructed and the particular design to employ. The Society recommended a reflector and particularly one designed by Thomas Grubb in the late 1840's.[4] Grubb had offered his design in reply to a statement by both the Royal Society and the British Association that it would be to the betterment of astronomy if there could be constructed in the Southern Hemisphere a telescope large enough to increase the number of images that could be detected in the sky without losing the high quality necessary for astronomical observation. When the telescope was up and tested, it was a subject for public ridicule. The telescope was easily vibrated and any disturbance to the tube—no mater how slight—caused the object of vision to dance all about the field of vision; the telescope was not placed within a wind-breaking dome but at the top of a building with a removable roof, thus adding to greater problems with imaging as a result of even the slightest breeze; and, finally, the mirrors used in the telescope had been installed with a protective coating, which when removed, required repolishing the mirrors; and the repolishing effort was highly unsatisfactory.[5] Despite the failure of this great effort, it became clear to designers and astronomers that mirrors did indeed have a great superiority over lenses.

The great advantage to the use of mirrors is that mirrors can eliminate the chromatic aberration that occurs with lenses. Just as importantly, mirrors reflect instead of absorbing ultraviolet light, and, as photography was now playing a more important role in astronomy, this second advantage was of extreme importance. Designers and astronomers began to place more emphasis on the construction of reflective telescopes so that today we find the reflecting telescope represents the largest telescopes to be found in observatories around the world.

Thus, the important differences between a reflecting and a refracting telescope have to do with the way in which light is collected in either instrument. The reflector depends upon the installation of a rather large and curved mirror of glass that is coated with a reflecting layer of aluminum. The mirror collects starlight and bounces it to other mirrors and eventually to a focus. The design in the refractor is markedly different. Here starlight is collected by large lenses that focus the light at the bottom of the telescope's tube. Design problems greatly limit the size of refractors.

In the cadiotropic telescopes, where there is a marriage of mirrors and lenses, the advantage to observers comes in the wider field that can be examined. The Schmidt Camera telescope is an example of a cadiotropic system. It not only allows observers a wider field for observation but it is also unaffected by the usual types of aberrations (like aperture or chromatic) which plague observers using other telescopes. This means the cadiotropic telescope can be used at very large aperture settings and still provide an extremely clear image. In one mirror systems, the field of vision will never be flat because the basic design necessitates a curved field of vision. This is indeed a problem but it can be resolved by placing a second mirror in such a way that the curvature is reduced or eliminated. This use of a second mirror for the flattening effect is known as the Schmidt-Cassegrain system. But one problem remains with the use of the additional mirror and that is the reduction in the size of the field that can be examined.

The fundamental idea of a Cassegrain system is that higher focal ratios, which allow astronomers very high resolution, can be effected by using a primary mirror with a hole in its center and a convex secondary mirror. The idea necessitates placing the mirrors so that the convex mirror reflects all light being received to the hole in the primary mirror. The light rays are, therefore, actually observed from behind the primary mirror. Because the secondary is a convex mirror, it reduces the amount of convergence from the light beam while it adds to the focal length available from the unique design, thereby producing the greater focal ratios that astronomers need for planetary and stellar observations.

Generally speaking, the larger the telescope, the greater the detail, but this rule only applies to a limited degree. At certain magnification ranges, the additional increases in magnification that can be brought about by more powerful eyepieces only continues until something called the "diffraction limit" is reached. This diffraction limit varies from telescope to telescope, and it depends solely on the size of the lenses or mirrors in use. Once the diffraction limit is reached, the image continues to magnify but now becomes a fuzzy ball of light giving less and less detail as it grows. The larger the telescope, the greater the amount of light it can collect, and the better the chances the observer has of learning more about the most distant and most faint objects in the sky, like the distant galaxies and quasars.

Right now, the largest telescope is owned by the Soviet Union. It is a 236-inch reflector. Anything that size has to be a reflector because the mirrors in reflectors have no severe size restrictions and therefore there is little danger of the telescopes causing annoying image distortions. Any Earth-based telescope, regardless of its size, has to face the disadvantages of Earth's atmosphere. Disturbance in the atmosphere adds to excessive jumping about of the images being focused upon as well as a fuzzing-out of the images. This fuzzing-out is generally referred to as *spread*. The amount of spread is always measured in arcseconds. Optical astronomers must tolerate image spreads anywhere from .50 to 2.5 arcseconds on the average night.

Thus, site selection is extremely important for telescope installations (Fig. 4-3). As a result, the great observatories around the world have been constructed at sites offering minimal atmospheric interference—sites like Kitt Peak, Arizona; Palomar Mountain, California; and Mauna Kea, Hawaii. The Mauna Kea Observatory is the site for the largest telescope observing the submillimeter range of the spectrum, allowing astronomers a new opportunity: observation of those interstellar gas clouds in which stars are formed.

Although early astronomers spent hours upon hours watching the sky, like Galileo or Herschel, modern astronomy has become a high technology science. Even optical astronomy depends on computers and special electronic instruments and displays. Direct vision astronomy is a very uneconomical enterprise today because it is very slow and very undependable. The astronomer's eye is given to illusion and tiring and so has been replaced by another eye—the camera's. Light captured by the great telescopes around the world is now directed onto photographic plates. Having stellar images captured on photographic plates is of tremendous advantage to an astronomer. He or she can be free to perform a multitude of tasks and yet never

Fig. 4-3. Modern observatories are located at very carefully selected sites. This is an aerial view of the Kitt Peak National Observatory in Arizona (Courtesy of National Optical Astronomy Observatories).

have to worry about missing anything happening in the sky. And, of course, a photographic print can be much better analyzed than an astronomer's sketch, for everything is caught in place as it might not be because of the blink of an eye, a loss of concentration, or an inability to sketch the field in time. The very special advantage of photographic plates is that they allow historical searches to compare the sky at different times. "Say, look at this light here," an astronomer might say as he or she pours over a print. "Where did it come from? Was it here yesterday? Last week?" A search of the files tells if it is new, has moved from some other position, or has actually been at its noted position for some time.

Here is where the Schmidt telescope is utilized (Fig. 4-4). It has been especially designed for astronomical photography. Its wide viewing capabilities allow astronomers to observe even the great nebulae that drift in distant space. Photographic processing has its drawbacks in terms of true color reproduction, but ingenious methods have been designed to compensate for the problems. These methods include masking techniques which help minimize the ranges in color or brightness that usually become lost when the photographs are made. In addition to special processing methods, astronomers also employ computers to help salvage faint details and eliminate discrepancies in contrast.

Fig. 4-4. The Schmidt telescope dome at the Mt. Wilson and Palomar observatories. The Schmidt telescope, or Schmidt camera, is an excellent instrument for photographic surveys (Courtesy of National Optical Astronomy Observatories).

Many images of the solar system, the galaxies, or the distant lights are "computer enhanced" or "color coded." Color is produced by visual light being transmitted at certain wavelengths, and computers can be programmed to detect these wavelengths and code them accordingly. Color coding is particularly important, for the color of the light collected from space tells much about its source. The better the color images, the easier it is for astronomers to analyze the data carried on the different wavelengths. Among the important facts about an object indicated by the color of its light, as you have learned, are temperature ranges, speed, and direction. And color coding compensates for the usual drawbacks of photographic processing.

Photography has played an extremely important role in optical astronomy since the last quarter of the nineteenth century. It has been, remains, and will be an important method of observation. Electron detectors have also become very popular in astronomy, and in some places they have replaced photographic equipment, but because emulsions are available in more convenient sizes than the detectors, photography is still the preferred means for analyzing celestial images over a wide field. Generally, astronomical photography is dependent upon spectroscopic plates specially designed for astronomical use and high-speed computers to make the necessary adjustments for spectral resolution problems that constantly occur.

All of this special technology that is brought together to enhance the field of optical astronomy indicates that this type of observation is still very important and will be a part of celestial investigations as long as astronomy is a science. The ground-based telescope will always be of importance and provide its own special convenience. Today, just as yesterday, the great observatories around the world continue to make extensive contributions to astronomy; and they will continue to add to the storehouse of knowledge about the solar system and the systems beyond it.

## MODERN OBSERVATORIES

Most of the more than 300 observatories scattered about the globe are part of university or government research facilities. More than fifty of them are located within the United States, and a half-dozen or so are orbiting around the Earth. These observatories contain the latest technological advancements for skywatching, as well as sophisticated photograph laboratories. They are generally constructed so that their telescopes are contained within domes that can be opened in one way or another to allow the telescope a window to the sky. The range of equipment necessary for a modern observatory ranges from the largest telescopes to electronic detectors and computers.

All modern astronomical observatories are located at very carefully selected sites. The main criteria are easy accessibility and limited atmospheric interference. What is particularly important is that the site offers a high number of viewing nights each year. A full 365 days of viewing is much more than most astronomers ever hope for, and generally most of them are satisfied with 250 clear nights per year. The sky above an observatory must be fairly free of "noise" because telescopes are highly sensitive instruments and they magnify not only distant stellar objects but any atmospheric turbulence.

The largest refracting telescopes are in the United States, France and Germany. Yerkes (Fig. 4-5) holds the record with a 40-inch refractor that is used mainly to accumulate spectra and positional data. Founded in 1897, the observatory was named for Charles Yerkes (1837-1905), an American financier from Philadelphia. Once a clerk in a grain commission house, Yerkes eventually became a broker and finally an investor who built his fortune by trading in corporate stocks. At the age of 44, he moved to Chicago where, in 1892, he offered the University of Chicago the financial backing to build an observatory at Williams Bay, Wisconsin.

Fig. 4-5. A close-up of the 40-inch great refractor at the Yerkes Observatory in Williams Bay, Wisconsin (Courtesy of Yerkes Observatory).

The Lick Observatory on Mt. Hamilton in California (Figs. 4-6 and 4-7) has the second largest (36-inch) refractor. Among the oldest American observatories, Lick dates back to 1889, and is named after the millionaire, J. Lick, who backed its construction and insisted that it should be a part of the University of California. Lick never lived to see the completion of the observatory, for he died at the age of 80 in 1876; but his body was exhumed and buried beneath the great telescope for which the Lick Observatory is famous. The administrative center for

Fig. 4-6. The 36-inch refractor at the Lick Observatory on Mt. Hamilton in California. This is second in size only to the 40-inch refractor at Yerkes (Courtesy of Lick Observatory).

the observatory is now located on the Santa Cruz campus of the university; the observatory itself is in a two-domed building on Mt. Hamilton. Astronomers assigned to the Lick Observatory have discovered thousands of double stars, determined the velocities of more than 3,000 stars, discovered four of Jupiter's moons, and continue to add to our knowledge of galaxies, stars, quasars, and the structure of the universe.

Like astronomers at all observatories, those working at Lick spend very little time looking through a telescope. Most of the time, the telescopes at the observatory are busy taking photographs rather than collecting light for a peeking astronomer, or they are actually analyzing the light they collect with the help of other instruments.

The third and fourth largest refracting telescopes are in Meudon, France (36-inch) and Potsdam, Germany (32.7-inch). Founded in 1670, the Meudon Observatory is much more than an observational center; it also serves as the central memory bank for Paris Observatory's Stellar Data Center, from which can be obtained spectral, radial velocity, and photometric data on more than one million stars. The Potsdam Observatory, originally conceived to further develop the use of spectroscopy and photography in astronomy, is one of the first observatories to have departed from traditional construction ideas which necessitated the close positioning of separate observational chambers or the placing of observing rooms in locations immediately east and west of the central dome.[6]

The fifth largest refracting telescope is located in Pittsburgh, Pennsylvania at the Allegheny Observatory. Founded in 1914, this observatory is equipped with a 30-inch refractor to add to planetary and stellar data. The observatory specializes in planetary studies and stellar positional and motional studies.

The very largest reflecting telescopes are located in the Soviet Union, the United States, Chile, and Australia (Figs. 4-8 through 4-11). The largest reflector at this writing is the 236-inch at the Special Astrophysical Observatory, Zelenchukskaya, Crimea. This is a relatively new observatory, completed in 1976; but its telescope has a mirror di-

Fig. 4-7. A view of the Lick Observatory from the west. This is one of the oldest observatories in the United States and dates back to 1899. Though the observatory is on Mt. Hamilton, the administrative center for the observatory is located on the Santa Cruz campus of the University of California (Courtesy of Lick Observatory).

Fig. 4-8. Aerial view of the Cerro Tololo Inter-American Observatory in Chile. This observatory is actually a part of the Kitt Peak National Observatory. Its 158-inch reflector is of the same design as that at Kitt Peak (Courtesy of National Optical Astronomy Observatories).

Fig. 4-9. The 4-meter telescope dome at the Kitt Peak National Observatory (Courtesy of National Optical Astronomy Observatories).

Fig. 4-10. The 4-meter telescope at the Cerro Tololo Inter-American Observatory in Chile (Courtesy of Kitt Peak National Observatory)

Fig. 4-11. The 200-inch Hale telescope, the second largest reflector in the world. The largest reflector is located at the Special Astrophysical Observatory in the Soviet Union (Courtesy of Hale Observatories).

ameter 36 inches larger than that in use of the next largest reflector. The light-gathering capabilities of a telescope this size give the Soviets a great advantage in the accumulation of data on quasistellar objects appearing to be stars but having large red shifts (the mysterious quasars that astronomers are so eager to learn about), and in the study of all distant members of the sky.

The second largest reflector in the world is located at the Hale Observatory at Palomar Mountain, California. There is another Hale Observatory at Mt. Wilson, California and this also has a relatively large (100-inch) reflector, though it is only half the size of the Palomar telescope. Both observatories are named for George Ellery Hale, who invented the spectroheliograph and also became known for his studies of solar vortices and magnetic fields. He also helped organize the Yerkes observatory. The observatory at Palomar Mountain only dates back to 1948 and works in joint programs with the Mt. Wilson observatory. The management and associations are somewhat confusing, for the Mt. Wilson Observatory is responsible to the Carnegie Institute and the Mt. Palomar observatory is responsible to the California Institute of Technology, yet they have a single director.[7]

The program at Mt. Palomar is designed mainly for the study of the fainter members of the sky and color analysis, and the program at Mt. Wilson is designed for solar studies. These observatories have made countless contributions to the great storehouse of astronomical data in computers around the world, from new and startling data about the Sun and stellar formation in general to the physical and chemical characteristics of all stellar objects. Some of the famous men in astronomy who have been associated in one way or another with the observatories include A.A. Michelson, E. Hubble, and W. Boade, whose backgrounds and accomplishments have already been listed in Chapter 1.

The third largest reflector is at Mt. Hopkins Observatory in Arizona, and this is a 177 inch (4.5 meter) telescope.

The Kitt Peak National Observatory in Arizona and the Cerro Tololo Inter-American Observatory in Chile are each equipped with a 158-inch reflector. These observatories are actually extensions of each other. The Kitt Peak Observatory, completed in 1959, was sponsored by the National Science Foundation and the Association of Universities for Research in Astronomy. Realizing that another observatory was necessary in order to survey the Southern Hemisphere, the National Science Foundation teamed with the National University of Chile in 1963 to construct the South American counterpart of Kitt Peak Observatory, and in 1962 the Cerro Tololo Observatory was founded. Both the Kitt Peak and Cerro Tololo Observatories are modernly equipped, with the Kitt Peak location maintaining about one dozen different telescopes (Fig. 4-13) and the Cerro Tololo Observatory maintaining about eight. The 158-inch reflector at Cerro Tololo is of the same design as the large reflector at Kitt Peak.

Fig. 4-12. This unusually designed building houses the McMath Solar Telescope and Vacuum Solar Telescope. The unusual design of the McMath telescope is important for achieving proper resolution for solar studies (Courtesy of National Optical Astronomical Laboratories).

The fifth largest reflector is the 153-incher at the Siding Spring Observatory in Australia. Founded in 1975, the observatory is a joint effort between the Royal Society and the Australian Academy of Sciences that began back in 1967. The Siding Spring reflector is currently the second largest telescope in the Southern Hemisphere, second, of course, to the Cerro Tololo reflector. At Siding Spring, there are extensive programs for the purpose of studying the Sun, celestial colors in general, and polarization.

Tables 4-1, 4-2, 4-3 and 4-4 provide summaries

Fig. 4-13. Close-up view of the Mayall 4-meter telescope at the Kitt Peak National Observatory (Courtesy of National Optical Astronomy Laboratories).

of the statistics relating to the world's largest telescopes and to the major observatories in the northern and southern hemispheres. The staffs and programs at these observatories keep the sky under continued surveillance. The research projects are numerous and are often carried out in joint efforts with teams from other countries. Scientists are not primarily concerned with the usual political boundaries that divide countries. Their search is a search for knowledge about what appears to be an infinite universe. They realize that a collective effort is needed to move toward an understanding of everything that lies in the sky.

It would at first seem as though land-based observatories will become useless in the space age, for they are dedicated not only to the oldest form of observation—visual light—but are also trapped on land where atmospheric problems wreak havoc with research. Just the opposite is true. The telescopes that they employ are highly sophisticated light gathering and analyzing instruments employing the latest in computer technology, solid-state light detectors, and photographic equipment. Visual astronomy programs have added greatly to our knowledge of the Sun and, therefore, also of stars in general. Visual astronomy has also helped in the compiling of significant data about spectra, star positions and diameters, variable stars, interstellar gases, and the physics and relationships of the other planets in our solar system. It has also played an important role in mapping out the Milky Way and nearby galaxies. Visual astronomy supplements, and is supplemented by, all the other fields of astronomy, from molecular to gamma ray.

**Table 4-I. Major Refracting Telescopes.**

| Location | Aperture Inches | Meters | First Year Of Operation |
|---|---|---|---|
| Yerkes Observatory, Williams Bay, Wisconsin | 40 | 1.02 | 1897 |
| Lick Observatory, Mt. Hamilton, Calif. | 36 | .91 | 1888 |
| Paris Observatory, Meuden, Franco | 32.7 | .83 | 1893 |
| Astrophysical Observatory, Potsdam, East Germany | 32 | .80 | 1899 |
| Allegheny Observatory, Pittsburgh, Pa. | 30 | .76 | 1914 |
| University of Paris, Nice, France | 30 | .76 | 1893 |

**Table 4-2. Major Reflecting Telescopes.**

| Location | Aperture | Inches Meters | First Year Of Operation |
|---|---|---|---|
| Special Astrophysical Observatory, Zelenchuksaya, Crimea, Soviet Union | 236 | 6.0 | 1976 |
| Palomar Observatory, Palomar Mountain, California | 200 | 5.1 | 1950 |
| Mt. Hopkins Observatory, Arizona | 177 | 4.5 | 1979 |
| Kitt Peak National Observatory, Arizona | 158 | 4.0 | 1973 |
| Cerro Tololo Inter-American Observatory, Chile | 158 | 4.0 | 1976 |
| Anglo-Australian Observatory, Siding Spring, Australia | 153 | 3.9 | 1975 |
| Mauna Kea Observatory, Hawaii (United Kingdom Infrared Telescope) | 150 | 3.8 | 1979 |
| European Southern Observatory, Cerro La Silla, Chile | 142 | 3.57 | 1975 |
| Mauna Kea Observatory, Hawaii | 126 | 3.2 | 1979 |
| Lick Observatory, Mt. Hamilton, California | 120 | 3.0 | 1957 |

**Table 4-3. Major Observatories in the United States.**

| Observatory | Location | Special Research |
|---|---|---|
| Arecibo | Puerto Rico | Planetary, interstellar |
| Carnegie | Washington, D.C. | Radio, solar, interstellar |
| Hale | Mt. Wilson, California<br>Palomar, California | Spectral, faint objects |
| Harvard | Massachusetts | Spectral, variables |
| Hawaii | Mauna Kea, Hawaii | Faint objects |
| Haystack Hill | Massachusetts | Planetary, radio |
| Jet Propulsion | Goldstone, California | Planetary, planetary satellites |
| Kitt Peak National | Tucson | Spectral, faint objects |
| Lick | Mt. Hamilton, California | Spectral, positional |
| Lowell | Flagstaff, Arizona | Spectral, positional |
| McDonald | Mt. Locke, Marfa, Texas | Spectral, planetary, interstellar, radio |
| Mt. Hopkins | Arizona | Planetary, positional, spectral |
| NASA | Mauna Kea, Hawaii | Infrared |
| NRA(National Radio Astronomy) | West Virginia | Radio, planetary, interstellar |
| Perkins | Flagstaff, Arizona | Spectral |
| Princeton | New Jersey | Spectral |
| Sacramento Peak | New Mexico | Solar |
| Sproul | Swarthmore, Pennsylvania | Positional |
| Stanford University | California | Lunar, planetary, radio, solar |
| U.S. Navy | Washington, D.C.<br>Flagstaff, Arizona | Planetary, motional, positional |
| University of Colorado | Boulder, Colorado | Radio, planetary, solar |
| University of Michigan | Ann Arbor, Michigan | Spectral |
| University of Virginia (Leander McCormick) | Virginia | Spectral, positional |
| University of Wisconsin | Madison, Wisconsin | Spectral, positional |
| Yale | Connecticut | Positional, spectral |
| Yerkes | Williams Bay, Wisconsin | Spectral |

Table 4-4. Other Observatories Around the World.

| Observatory | Location | Special Reasearch |
|---|---|---|
| Algonquin | Ontario, Canada | Interstellar |
| Arcetri Royal | Italy | Solar |
| Baja California | San Pedro Martir | Spectral, faint objects |
| Basel | Metzerlen, Switzerland | Solar, spectral, faint objects |
| Bonn | Hoher List, Germany | Interstellar, spectral, positional |
| Boyden | Mazelspoort, South Africa | Spectral, positional |
| Burakan | Armenia | Faint objects |
| Cambridge | England | Spectral, solar |
| Cape of Good Hope | South Africa | Spectral, positional |
| Carnegie | Las Campanes, Chile | Faint objects |
| Berlin | Babelsberg, Germany | Positional, spectral |
| CERGA | Grasse, France | Positional, lunar, satellite, infrared |
| Cerro Tololo | Chile | Spectral, planetary, faint objects |
| Copenhagen | Brorfelde, Denmark | Spectral |
| Cordoba | Bosque Alegre, Argentina | Spectral |
| Crimean Astrophysical | Soviet Union | Interstellar, solar, planetary, faint objects |
| CSIRO | Parks, Australia | Spectral |
| David Dunlap | Toronto, Canada | Spectral |
| Dominion Astrophysical | Victoria, Canada | Spectral |
| Edinburgh Royal | Scotland | Positional, spectral |
| European Southern | La Silla, Chile | Spectral, faint objects |
| Hamburg | Bergedirf, Germany | Solar, positional, faint objects |
| Haute-Province | St. Michele, France | Spectral, faint objects |
| Helwan | Cairo, Kottomia, Egypt | Spectral, faint objects |
| Jodrell Bank | England | Solar, planetary, radio, interstellar |
| Geneva | Jungfraujoch, Switzerland | Spectral |
| La Plata | La Plata, Argentina | Spectral, positional |
| Mt. Stromlo | Canberra, Australia | Spectral, positional |
| Mullard | Cambridge, England | Solar, planetary, radio |
| Nice | France | Stellar, positional |
| Okayama | Japan | Solar, faint objects |
| Ondrejov | Czechoslovakia | Spectral, positional, solar |
| Padua | Asiago, Italy | Spectral, faint objects |
| Paris | Meudon, France | Spectral, planetary, solar |
| Peking | Sha-ho, Mi-yun, Hsing-lung, China | Positional, solar, spectral, radio |
| Penticon | British Columbia | Spectral, interstellar, solar, cosmic |
| Pic du Midi | Pyrenean Mountains, France | Planetary, solar |
| Pulkovoo | Leningrad | Spectral, positional, planetary, solar, radio |
| Purple Mountains | Nanking, China | Solar, positional, stellar |
| Royal Greenwich | Herstmonceux, England; Canary Islands | Positional, solar, spectral, stellar |
| Schwarzchild | Tautenburg, Germany | Stellar, galactic |
| Shanghai | China | Positional |
| Siding Springs | Australia | Spectral, faint objects |
| Skalnate Pleso | Czechoslovakia | Spectral, solar, positional, stellar |
| Stockholm | Saltsjobaden, Sweden | Positional, spectral |
| Sydney University | Narrabri, Australia | Stellar, radio |
| Radcliffe | Sutherland, South Africa | Spectral, stellar |
| Tokyo | Mitaka, Japan | Solar, spectral, radio, positional |
| Tonantzintla | Mexico | Spectral, stellar, galactic |
| University of Leiden | Dwingeloo & Westerbork Netherlands | Interstellar, radio |
| University of Paris | France | Interstellar |
| Uppsala | Sweden | Stellar, galactic |
| Vatican | Castel Gandolfo | Spectral, positional |
| Yale-Columbia | El Leoncito, Argentina | Positional, spectral |
| Yunnan | Kunming, China | Positional, solar |
| Zelenchukskaya | Soviet Union | Spectral, radio, stellar, galactic |
| Zurich | Arosa, Switzerland | Solar |

Fig. 4-14. A 120-inch mirror on a grinding machine at the Lick Observatory (Courtesy of Lick Observatory).

A great deal of data still needs to be collected about the Sun and the Moon. Astrophysicists are constantly devising newer and more fascinating ways in which to learn more about this star and this satellite. The Moon holds its mysteries. It is much too big to have been formed as most natural satellites; that is, from a piece of a planet that was hurled into space. Our Moon is much too large to once have been a chunk of Earth. It was probably a planet in orbit about the Sun millions or billions of years ago, but its erratic orbit brought it too close to Earth's gravitational field and it was captured in Earth's orbit. The other moons that probably orbited Earth at this time eventually collided with the Moon and so now Earth has only one natural satellite. As for the Sun, there is so much happening at and below the atmosphere that its mysteries will not be unraveled for hundreds of years, if ever.

Observational programs designed to study the Sun must differ somewhat from those designed to study other celestial objects because the Sun is not only exceptionally bright, but it is also extremely close. You can almost guess what these two characteristics must do to observational efforts. The extreme brightness means that the Sun cannot be observed directly and the extreme heat means that Earth's atmosphere is going to be so disturbed that it will hardly allow daytime viewing. Additionally, the extreme solar heat causes image distortion.

There are generally two types of telescopes that observatories employ to study the Sun: disk telescopes and coronagraphs. The most efficient disk telescope is a solar tower designed with two flat mirrors used to catch sunlight and reflect it into a refracting or compound reflecting telescope.[8] This results in a solar image produced near ground level where the radiation can be collected efficiently. The coronagraph is designed to entirely eliminate the problems resulting from scattered light, so that astronomers can observe the Sun's corona (which is very close to the disk). Astronomers are always particularly interested in the Sun because it is the closest star that can be observed. The next closest star is some 4.3 light years away. This means that if you were able to travel at the speed of light, a speed which would take you around the Earth's equator at a little better than seven times a second, it would still take you more than four years to reach that star. Beyond Alpha Centauri, there are only six stars less than 10 light years away and these are Barnard's Star (6 light years), Wolf 359 (7.7 light years), Luyton 726-8 (7.9 light years), Lalande 21185 (8.2 light years), Sirius (8.7 light years), and Ross 154 (9.3 light years). Most of the stars in the sky are thousands of light years away. It is no wonder, then, that astronomers must depend upon their studies of the Sun to learn about the stars.

Observatory research programs have also made significant contributions to our understanding of novae and supernovae. Analyses at the invisible wavelengths have helped, but there is still a great deal to be learned about novae and supernovae. Acquiring of that knowledge is dependent upon the use of telescopes with wide fields of view. Novae and supernovae are so far away that they are, at the very best, at about the 16th magnitude, so there is a need

**Table 4-5. Nearby Stars.**

| Name | Approximate Distance (parsecs) | Proper Motion (Annual, sec. of arc) |
|---|---|---|
| Alpha Centauri | 1.31 | 3.68 |
| Proxima Centauri | 1.31 | 3.68 |
| Barnard's Star | 1.83 | 10.32 |
| Wolf 359 | 2.32 | 4.80 |
| Lalande 21185 | 2.50 | 4.78 |
| Sirius (A) | 2.67 | 1.32 |
| Luyten 726-8 (A) | 2.70 | 3.35 |
| Ross 154 | 2.90 | 0.70 |
| Ross 248 | 3.15 | 1.60 |
| Epsilon Eridani | 3.28 | 0.97 |

A parsec is equivalent to 3.26 light years or 19.2 trillion miles (30.9 trillion kilometers). Distances to stars are subject to constant revision. Note that some sources establish the distance to Luyten to be closer than Lalande. Also note that *proper motion* means what appears to be the angular motion of a star (on an annual basis) on the celestial sphere as it moves perpendicular to an observer's line of sight. Nearby stars are not always the brightest stars. In fact, only two stars in this table would make a list of the 25 brightest stars and these would be Sirius and the Centauri stars.

for very high-powered photographic instruments that can capture fields and objects well beyond our Milky Way Galaxy.

Novae are stars which suddenly grow in brightness. This brightness can increase by a factor of about 4,000 in a span of time measuring no more than a few days. They are dim little specks to an observer—or on a photographic plate—and appear to be both within and without the Milky Way Galaxy. But within the Milky Way system, there are no more than a dozen or so novae spotted each year, and these generally dim again within months or weeks after the initial increase in brightness. Their emission lines reveal that they are very hot; their spectral data constantly changes and is very, very complex. Generally, novae seem to occur as part of binaries. The distance of each binary star from one another seems to be roughly equivalent to the actual size of the stars, and the chemical and physical characteristics of the stars are such that one binary is very hot and very compact but the other is cool and somewhat extended. The physics of the situation creates a condition whereby gaseous accretions by the compact (dwarf) star cause it to emit

Fig. 4-15. Equitorial star trails (Courtesy of Lick Observatory).

extra light. If there is continuous transfer of material from the extended (subgiant) star, there is the possibility that the mass of the dwarf reaches a point that assures the dwarf of nothing less than a cataclysmic fate. The dwarf collapses into a neutron star and releases such an incredible amount of energy that it becomes a supernova. A supernova is an exploding star that reaches a brightness that may be 10 billion times greater than normal. These supernova explosions are relatively rare. In a star population of, perhaps, a billion stars, astronomers predict that, at best, only one supernova explosion occurs every couple of dozen years. In the case of supernovae, spectral observations show extremely high amounts of gas thrust into space at thousands of miles per second. Physicists guess that supernovae result whenever a star about four times as massive as the Sun dies (meaning its core collapses, resulting in a catastrophic explosion).

Now, while the great observatories play important roles in solar, solar system, and stellar observations, they are also engaged in systematically surveying the sky by photographic means in order to map the sky, catalog celestial objects, and investigate variable stars. Variable stars are stars which have detectable changes in brightness. The factor of change here may be thousands of times what had been previously observed. There can be any number of reasons for these changes in brightness: varying atmospheric conditions, motions of the stars in relationship to other objects between them and Earth, or an actual physical change in the makeup of the star. Variable stars are generally classified as long-period, short-period, or irregular-period. Long-period variables are the most common, and are expected to be red giants; these may have periods of three months to three years but some may possibly have periods extending for hundreds or thousands of years. The short-period variables are white high-luminosity stars that have periods from less than a day to up to three months. The irregular variables are unpredictable and are usually red or yellow dwarf stars; they are also highly eruptive stars and may possibly be novae or supernovae. Studies of variables, or of star variations in general, adds greatly to cosmological models and assists astronomers in determining the dimensions of the Galaxy and distances within the Galaxy. (Astronomical statistics and measurements are always under constant revision.)

Earth-based observations of the transit of astronomical objects is also of considerable importance in astronomy. "Transit" means the apparent movement of a planet across the Sun, a star across the local meridian, or other celestial objects across some point of reference convenient to the observer. Planetary transits are particularly important for determining the exact position of a planet or the exact details of its orbital characteristics. Stellar transits are important for the compilation of positional data, for navigation, and for cosmic timekeeping.

Though it remains secondary to no other branch of astronomy, visual astronomy only represents one of the many ways of investigating the universe. In coming chapters, we will investigate these other ways.

# Chapter 5
# Radio Astronomy

RADIO ASTRONOMY IS CONCERNED WITH COMMUNications from outer space that can be detected by reading electromagnetic radiation at wavelengths from roughly 1 mm to 30 m. The first scientist to detect radio emissions from distant celestial objects was a young engineer by the name of Karl Jansky (1905-1950). Jansky was employed by the Bell Telephone Laboratories in New Jersey. He joined them in 1928 and was assigned to troubleshoot problems with short wave transmissions. Ship-to-shore radio communications were continually hampered by noise (static interference), and Bell Labs wanted to find the cause of the interference. Jansky began to make progress in the summer of 1931 when he used a rotatable antenna array, which he had designed and built, to record the intensity of the static and the direction from which it was coming. He operated at a wavelength of 14.6 meters. He picked up a lot of noise that was clearly caused by weather conditions, but some of the static in transmissions was actually coming from somewhere out in space. And when he turned his antenna toward the very center of the Galaxy, he detected a steady, hissing static that was more intense than usual. He soon came to the conclusion that space itself, or rather objects that existed in deep space, were constantly transmitting radiation at this wavelength. Jansky had no way of knowing it then, but those strange signals which he was picking up were caused by elementary particles moving through galactic magnetic fields. These particles are electrons which have a negative charge of electricity equal to about $1.602 \times 10^{-19}$ coulomb and are roughly $1/836$ the mass of a proton.

From the New York Times, May 16, 1933:

> Electrical energy in the form of radio waves, which scientists believe come from a point so remote in space that it requires between 20,000 and 40,000 light years for the waves to reach the earth, was heard last night by radio listeners throughout the United States. It was the first such experiment ever carried out. The sound, generated by

Fig. 5-1. Karl Jansky checking readouts of transmission data. He was the first to detect radio waves in space as a result of his troubleshooting of ship-to-shore radio problems for Bell Laboratories (Courtesy of AIP Niels Bohr Library).

the waves arriving at a super-sensitive receiving set operated by Dr. Karl G. Jansky, research engineer of the Bell Telephone Laboratories' experimental station at Holmdel, N.J., sounded like steam escaping from a radiator. Wires carried the sound from the New Jersey receiving station to the WJZ coast-to-coast network.

Dr. Jansky, speaking of his work carried on secretly for more than a year, the results of which were recently reported in THE NEW YORK TIMES, said an immense amount of electrical power would be necessary to transmit waves over such distances. Some of the stars, however, have been found to radiate as much as 500 sextillion horsepower, he added.

Dr. Jansky was introduced by O.H. Caldwell, former Federal Radio Commissioner, who explained how the research engineer, using an antenna rotated by motors, determined the point in the sky from which the waves apparently arrive through space. The rotation of the earth on its axis causes the waves to strike the earth at different angles, depending upon the time of day and the season of the year. By carefully checking the gathered data it was discovered that the waves were arriving from the region of the constellation of Sagittarius (The Archer).

"Our observations show definitely," said Dr. Jansky, "that the maximum of hiss comes from somewhere on the celestial meridian designated by astronomers as eighteen hours, right ascension, and we are sure of this to a few degrees. It happened that my observations were made on 146 meters, or a frequency of 20,600 kilocycles, although energy also seems to come in on wave-lengths both above and below that part of the wave band. I expect that interstellar static impulses will be found all up and down the radio spectrum, probably increasing in the high frequency portions."

Dr. Jansky said the directions of travel of the waves in space "seems to confirm Dr. Harlow Shapley's calculations that the energy probably comes from the center of gravity of our Galaxy. You can realize that if all the stars in one galaxy were sending out waves such as we detect with the instruments at Holmdel, they might naturally appear to come from the center of gravity of the system."

The solar system has been estimated to be about 250,000 light years across with the earth situated approximately 40,000 light years from one edge.

Some years later, in 1940, another engineer by the name of Grote Reber confirmed Jansky's findings. Reber was always fascinated by Jansky's discovery and even as far back as 1932 he was attempting to detect interstellar radio waves with his own shortwave radio. But these first attempts were of no consequence; but Reber was undaunted. Five years later he constructed an antenna about 9.4 meters in diameter that was, in fact, the very first radio telescope. In 1940, he was able to confirm Jansky's findings. But he did something more. Turning his parabolic reflector toward the solar region, he found that there were also radio signals coming from the Sun. Astronomers were now beginning to realize that there were other means, other than optical, for observing the solar system and objects beyond.

Reber led the way. During the Second World War, he took the first preliminary reading of the radio sky, successfully mapping high-frequency shortwave areas. In 1951, he designed and constructed a new radio telescope in Hawaii, this one to be used for low-frequency readings; and afterward he continued to use his genius to collect, analyze and map radio sources.

Radio astronomy has come a long way since Jansky first detected those strange signals from the center of the Galaxy. The major instrument in the

**Table 5-I. Major Radio Telescopes.**

| Location | Reflector Size (meters) |
|---|---|
| Algonquin Radio Observatory, Lake Traverse, Ontario | 100 |
| California Institute of Technology | 40<br>10 |
| Crimean Astrophysical Observatory, Crimea, Soviet Union | 22 |
| CSIRO, Parkes, Australia | 64 |
| Haystack Observatory, Westford, Massachusetts | 37 |
| Jet Propulsion Laboratory Goldstone, California | 64 |
| Max Planck Institute for Radio Astronomy, Effelsberg, West Germany | 100 |
| National Astronomy and Ionosphere Center, Arecibo, Puerto Rico | 305 |
| National Radio Astronomy Observatory, Kitt Peak, Arizona | 11 |
| National Radio Astronomy Observatory, Green Bank, West Virginia | 43<br>91 |
| Nuffield Radio Astronomy Laboratory, Goldstone, California | 76 |
| Observatory of Paris, Nancay, France | 40 × 200 |
| Onsala Observatory, Gothenburg, Sweden | 20 |
| Special Astrophysical Observatory, Zelenchuksaya, Soviet Union | 10 × 1885 |
| Tata Institute, Ootacamund, India | 30 × 259 |
| University of Amherst, Massachusetts | 14 |

Fig. 5-2. The Jodrell Bank Radio Telescope. The sky is so radioactive that astronomers need only point their telescopes in any direction to begin receiving signals (Courtesy of AIP Niels Bohr Library).

Fig. 5-3. The 64 meter antenna in Goldstone, California (Courtesy of NASA).

advancement of this field has been, of course, the radio telescope. Basically, it consists of a reflector, an electronic receiver and a display device, all designed to sense and measure the spacial signals coming in on the radio wavelengths of the electromagnetic spectrum. The reflector serves as a surface for collecting and focusing the radiation; the electronic receiver is a sophisticated device for both amplification and detection; and the display is a video device for giving observable form to the electromagnetic data. Whereas Jansky actually listened to cosmic radio signals, astronomers today study images of those signals. These image displays are important in helping astronomers with the three W's of electromagnetic radiation: What are the signals? Where are the signals coming from? What are their sizes and shapes? Astronomers always have the option of listening to radio signals from space if they wish, but there is little practical advantage to eavesdropping on the universe.

There are some unique problems in radio astronomy to do with the range at which observations can actually be made. There are serious restrictions, some natural, some contrived. To begin with, there are politically imposed limitations which have resulted because of the needs for standards that allow broadcasters freedom to use certain wavelengths; if radio astronomers tried to use these broadcasting

Fig. 5-4. The steerable radio dish at Jodrell Bank, Macclesfield, Cheshire, U.K. This radio dish, which is the principle instrument for the Nuffield Radio Astronomy Laboratories, has been in operation since 1957 (Courtesy of Nuffield Radio Astronomy Laboratories).

**Table 5-2. Radio Astronomy.**

| System | Major Sources |
|---|---|
| Solar | Sun, Jupiter |
| Galaxy | Supernovae remnants, pulsars, masers, HI regions, HII regions, galactic center |
| Extragalactic | Some spiral galaxies, radio galaxies, quasars |

wavelengths, radio signals from space would be completely buried by public and private broadcasting efforts. Secondly, there are the natural restrictions imposed by atmospheric conditions around Earth (Table 5-3).

Earth's ionosphere is that part of the atmosphere which fills the sky from about 30 miles (50 kilometers) to 300 miles (500 kilometers) or more above the surface of the Earth. It is a field of electrically-charged particles which transmit radio waves around Earth. Longer wavelength radiation coming from outerspace is reflected back into space rather than allowed to penetrate the atmosphere and make it to the surface where stand the radio telescopes poised to collect the communications. At the very short wavelengths, nature has provided an additional complication. Carbon monoxide and other gases in the atmosphere absorb radiation at these shorter wavelengths. Thus, radio telescopes are limited to observations at wavelengths less than 30 m and not more than 1 m for these reasons alone.

## RADIO SPACE

We actually live in a sea of radio radiation. Radio emissions and reflections come from just about everything that exists. For instance, as you read these words you are broadcasting radio signals to the people around you, or off the walls and objects about you. However, you cannot be conscious of the signals you are receiving or sending, and you are indeed receiving them as well as sending them. You cannot be conscious of them because the emissions or reflections are not very strong. (That is, perhaps you cannot be conscious of them; it may be that much of what is attributed to psychic phenomena may actually be the result of certain individuals be-

**Table 5-3. Earth's Atmospheric Layers.**

| Region | Description |
|---|---|
| Troposphere | Extends 7 to 10 miles (11 to 16 kilometers) from Earth's surface. There are marked decreases in temperature as altitude increases. |
| Stratosphere | Extends 7 to 31 miles (11 to 50 kilometers). Latitude, season, and weather all affect the actual range of the stratosphere. There is little temperature change throughout the region. |
| Mesophere | Extends from the top of the stratosphere to 50 miles (80 kilometers). Temperature here decreased with altitude. |
| Thermosphere | Extends into outerspace and begins about 50 miles (80 kilometers) above the Earth's surface. Temperature here increases with altitude. |
| Ionosphere | Extends anywhere from 30 miles (50 kilometers) to 300 miles above the surface. Its physical characteristics allow extended transmissions of radio waves about the Earth. |
| Magnetosphere | Extends for thousands of miles; the upper atmospheric region. |

Earth's atmosphere protects the planet and the life on it by absorbing or reducing cosmic and other radiation and by creating a friction furnace which burns away most of the solid objects that hurtle toward the surface.

ing able to sense and interpret (consciously or unconsciously) invisible radiations. This idea is not so extraordinary as it may seem, for, after all, we are able to sense visual radiation, and this is no more than electromagnetic radiation at certain wavelengths.

The sky is so radioactive that radio astronomers need only point their telescopes skyward to pick up signals. Just about everything moving in the sky is a source of radio radiation—and everything up there moves. Signals are constantly being received not only from within our solar system and within our Galaxy but from other galaxies within the Local Group and beyond. These signals can be detected around the clock. There are no special time restrictions on radio astronomers; they can observe from sunrise to sunset. While signals come from all over and at all times, they do not all come with equal intensity.

In their attempts to detect and measure radio emissions from space, astronomers have developed highly detailed maps of the sky. These maps do not in themselves indicate specific sources of radio signals; to find these sources there must be a marriage of the observational and radio branches of astronomy. Using actual photographs of the sky for the necessary positional notations, and radio images of the sky for directional aid, astronomers can now locate the transmitting object or field. But a great deal of the radiation seems to have no source at all. Even though the radio maps indicate signals are coming from certain coordinates, the optical photographs show nothing at all in existence at the locations given. This may simply mean that there are galaxies or other objects much too far away to be photographed but not so far away that their radio emissions cannot be detected. If this is indeed the answer then it is quite apparent how much further into the universe radio astronomy can bring us. And because of radio astronomers' great reach, cosmologists hope that it may lead them to where lie the answers to the great questions of how and when—and maybe why—the universe began.

In the future, radio astronomy may take to space. The Jet Propulsion Laboratory conducted some very successful tests in the summer of 1986 with a ground-based telescope and the 4.8 NASA Tracking and Data Satellite. Objectives of the tests were quasars PKS 1730-130, PKS 1741-038, and PKS 1510-089. Researchers at Jet were highly enthusiastic that as a result of their efforts both the range and resolution to be expected from radio research will soon be far more advanced than it has ever been. At this writing, NASA and the European Space Agency are considering launching a satellite (QUASAT) which will, in effect, be a 15-meter antenna designed for interferometric observation in unison with ground-based radio research.

## THE SUN

The Sun, on which each and every body in the solar system is dependent, is a tremendous source for all types of electromagnetic radiation. As might be expected, the greater percent of the radiation coming from this star is in the 400 to 800 mm range of the electromagnetic spectrum, which is the range of visual light. This is by no means the only range of wavelengths at which the Sun communicates with Earth; the wavelengths extend from one end of the spectrum to the other. As well as being the greatest source of light to Earth, the Sun is also the greatest source of radio emissions.

The Sun is tremendously large and bright compared to everything else we can observe in the sky, but this is only due to its proximity to Earth. The Sun, as far as can be detected, is simply an average star. Generally speaking, there is nothing especially unusual about its physics. It is of average temperature for a star, of average brightness, and of average age. It is spectacular, however, in that it governs, so to speak, a solar system containing one-life giving planet; and without the Sun, life could never have occurred on Earth or been sustained as long as it has been. (Some might take exception to this statement for it is always possible that the Earth originated before the Sun and was eventually stolen from another star or another gravitational field by the strength of the Sun's gravitation, much as the Moon might once have been on its own orbit about the Sun before the Earth captured it. That the Earth strayed into the Sun's orbit, however, is highly unlikely. It appears much more likely that Earth and all the planets were born from fragments of the same great nebula that produced the Sun.)

It is not just simply the fact that the Earth is heated by the Sun that explains its fertility. If this were so, then many of the other planets in the solar system would also be sustaining some type of life form right now. Life on Earth has occurred because of a number of very special conditions, not the least of which is its mean orbit from the Sun (149.6 million kilometers). Any nearer or further from the Sun, and Earth might be more like a Venus or a Mars. Its atmosphere would never have evolved in quite the same way that it has. To be truthful, there is little certainty about the evolution of the Earth's atmosphere, but there are some very logical theories. One explains that due to the very low temperatures (roughly 25 degrees celsius), during the accretion stages of its evolution, Earth would have been secured in an aqueous envelope because of the water vapor resulting from the impact of planetesimals. If the surface temperatures were much more than 25 degrees, then there would have been a mixing of these vapors with the carbon dioxide in the atmosphere; the result would be a trapping of infrared radiation and the heating of Earth to temperatures comparable to those of Venus, and life could never have gotten its chance. By some special accident or special design, the way in which the condensing vapors combined with atmospheric elements resulted in a considerable reduction of the carbon dioxide in the atmosphere. This meant that there could be the necessary water clouds to filter the Sun's burning rays, thereby providing the cooler temperatures necessary to support life.[2]

Thus, the spectacular aspect of the Sun is not simply that it provides heat, but that its gravitational interactions have trapped the Earth in an orbit that has allowed it, since millions or billions of years ago, to be in a position to receive exactly the right amount of heat.

Close observation of the Sun indicates that the outer part of its atmosphere is a gaseous envelope called the corona. As you learned in a previous chapter, the corona is an extremely high temperature zone, even for a star of the Sun's classification. Temperatures in the corona are in the millions of degrees. This causes the Sun's atmosphere to actually be hotter than the Sun's surface. And, as you know, the reason for this is the way in which solar energy is produced and transmitted. Energy, of course, takes many forms. There can be heat energy, electrical energy, nuclear energy, and many other types, not the least of which is mechanical energy. This last form of energy is actually created by the Sun when convective processes put just enough pressure on internal gases to force them toward the surface. This creation of mechanical energy within the Sun is extremely important to astronomers because mechanical energy is much more easily transmitted through those invisible walls of subatomic particles that would ordinarily hamper radiant or convective forms of energy. The production of mechanical energy, then, allows astronomers to "see" below the surface of the Sun simply by examining this energy production. Thus, by studying the corona, astronomers can "see" into the sun's interior and the internal convective processes produces radiation at all wavelengths.

Besides the continuous transmissions from the Sun's corona, there are also radio emissions with the occurrences of solar flares. Associated with those great gas clouds called prominences, solar flares are those bright clouds called prominences, solar flares are those bright clouds that come across as black lines (filaments) when viewed with the disk of the Sun as a backdrop. Usually floating somewhere between the corona and chromosphere, prominences can sometimes be observed in unusual forms and at unusual speeds, moving either to or away from the surface of the Sun because of local explosions (flares).

(The atmosphere and the surface of the Sun have been discussed in Chapter 3, but we must repeat some of the information here in order to relate the solar phenomena with radio emissions.)

These local solar explosions called flares may last only for a few minutes or they may last for a few hours. But brief or not, they are always extremely violent eruptions. Though flares cannot be detected in white light, they are easily observed in the red light of hydrogen or else detected by shortwave failures in the daylight hemisphere of the

Earth. These powerful solar storms blow high-energy electrons and protons into the atmosphere where they cause other atmospheric elements to begin radiating. The frequency at which the radiation is transmitted is referred to as the plasma frequency, and, as it depends upon the density of the medium through which it must pass, it is never constant. Radiation is also produced within the Sun's magnetic field where trapped electrons, trying to escape at the speed of light, radiate easily measurable amounts of radio signals.

## THE PLANETS

The planets in our solar system are numerous. They number in the tens of thousands and possibly in the hundreds of thousands. This is because the word "planet," in its most basic definition, refers to anything, except meteors or comets, that has its primary orbit around the Sun. We commonly think of the nine major planets—Mercury, Venus, Earth, Mars, Jupiter, Saturn, Uranus, Neptune, and Pluto—as the only planets in our solar system, but there are indeed others: the so far uncounted number of asteroids that orbit the Sun, primarily between the orbits of Mars and Jupiter. There may be additional planets the size of Mercury or Venus orbiting beyond Neptune or Pluto but, if so, they remain undetected. Quite contrary to belief, much of our solar system is still unknown.

The reason that much of the solar system is still unmapped is the way in which planetary orbits and the speed of planets result in very complicated relative motions and positions. This allows for relatively large worlds to remain hidden from view. For instance, until 1986 when Voyager 2 orbited Uranus, there were ten natural satellites that remained unknown. Quite possibly, when Voyager 2 visits Neptune about the time this book is published, more worlds will be found. The relative motions of all bodies can keep even stars the size of the Sun undetected. Everything in the solar system moves according to basic laws. Johannes Kepler gave us the three laws which govern the motion of planets. These may be stated as follows:

- Planets move in elliptical paths, and the Sun is one of the focal points of the ellipse.
- Each planet revolves around the Sun so that the line which connects the planet and the Sun sweeps out equal areas in equal times. Thus, a planet's velocity decreases as its distance from the Sun increases.
- The squares of the sidereal periods of two planets will be proportional to their mean distances from the sun.

Radio astronomy has helped to answer some of the more penetrating questions about the major planets. Take, for instance, the core of the planet Mercury, the closest planet to the Sun and possibly the second smallest of the nine major planets (if distant Pluto is as small as astronomers currently believe).

Orbiting roughly 46 million to 70 million kilometers from the Sun, Mercury is only visible by telscope during the day. With an equatorial diameter of about 4,880 kilometers, it is only one-third the size of Earth, and because of its size and its surface characteristics it can easily be mistaken for the Moon. Though it was not until the Mariner probes of the 1970s that very much was learned about the planet, its existence has been known for five thousand years. Its orbital pattern has always intrigued astronomers, mainly because it is not as circular an orbit as seven of the other major planets have. (Pluto, like Mercury, also has a highly eccentric orbit). For the longest time, it was believed that Mercury orbited the Sun in such a way that one of its faces never looked at the Sun. It took radio surveys of the planet to prove this theory false. If indeed one side of the planet never faced the Sun, then its surface temperatures should be drastically cooler on the other side of the planet, this was never detected in radio studies. Both sides of the planet have proved to be extremely hot.

Venus has also intrigued astronomers, mainly because it's a planet that appears to like a great deal of privacy. Most of what has been compiled about this planet, beyond the usual and easily observable phenomena related to its unusual brightness and position, has only been learned from ingenious but painstaking efforts, including the use of radio and radar and spacecraft. Its exceptional dense cloud covering has always been like a soft but very thick curtain allowing no view to the surface. Even its retrograde rotation could not be discovered by simple optical observations; astronomers have had to detect changes in the frequency of radar waves reflected from approaching or retreating edges of the planet in order to discover the retrogradation. And if it were not for the efforts of radio astronomers, there would now be very little compiled about the atmospheric conditions of the planet or its roughly-complexioned surface.

It was radio astronomers who first detected the extremely high temperatures on Venus, temperatures that might reach as much as 10,000 degrees Celsius. This was an unexpected discovery. Remember, Mercury is far closer to the Sun than Venus and so it would be expected to be hotter, but the cloud covering on Venus has created a greenhouse effect that leaves the planet at least twice as hot as Mercury. Let's take time to review how that greenhouse occurs. The dense cloud covering traps infrared rays and the rays not only heat the surrounding atmosphere but the surface of the planet as well. Sunlight penetrating the clouds is reflected back into the clouds again, this time at infrared wavelengths. And because of the planet's predominantly carbon dioxide atmosphere, the infrared radiation is reflected once again back to the surface. Thus, the area between the surface and the clouds continues to heat up. The heat increases but can never escape. The result is a planet that may never be cool enough to allow life. (How lucky we are that the Earth's atmosphere developed in a very different way than the atmosphere of Venus and that it orbits at just the right distance from the Sun to allow stable temperatures!)

While radio studies of Mercury and Venus have been important in verifying or adding to the data collected by optical and infrared researchers, they have been of little value in the case of the planet Mars, except, of course, to occasionally be an argument for the possible existence of highly advanced civilizations existing on or under the planet's surface. Mars has always been a newsmaker and a popular environment in science fiction because it has some strange surface characteristics. It has polar regions like Earth, seasons like Earth, vast gray-green areas that some argue may be patches of vegetation, and extensive dry beds that past astronomers and journalists have described as possible canals. But Viking probes have turned up no evidence that Mars is a life-supporting system, and its radio emissions have never amounted to much since they were first detected in the late 1950s.

The first planet beyond Mars, however, is a source of extremely strong radio emissions. This is the giant planet Jupiter and the fifth planet from the Sun. Orbiting at an average distance of about 483 million miles (778 million kilometers) from the Sun, this is a very gaseous planet very much unlike Earth.

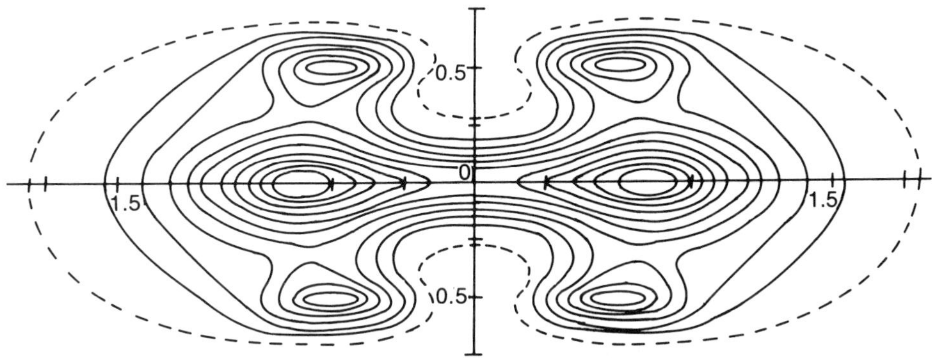


Fig. 5-5. This diagram illustrates the contours of radio emissions that come from Jupiter and appear to originate from the Van Allen belts (Courtesy of McGraw-Hill Encylopedia of Astronomy, Sybil B. Parker, Editor, 1982).

While Earth, Venus, Mercury, and Mars are rocky planets, Jupiter is a liquid planet with an atmosphere consisting mainly of hydrogen and helium. It has no solid surface. Radio radiation was first detected on the planet in 1955 and the reason for the intense emissions is still to be determined. The only object in the sky that is at present a greater source of radio emissions is the Sun, so the Jovian emissions — so far away and from a world much smaller than the Sun, yet so intense — are quite a puzzle. Adding to the puzzle is the fact that the radio emissions seem to be generated from an area that takes up less than 3 percent of the planet. What can possibly be generating the radiation? Astronomers still seek the answer.

As you advance further into the solar system and to the first planet beyond Jupiter, you come to Saturn. Like Jupiter, it is also a liquid planet with an atmosphere of mainly hydrogen and helium. However, compared to Jupiter, Saturn is a relatively weak source of radio emissions, which only adds to the mystery of the Jupiter emissions. Why should these two giant planets, so much alike in composition and size, be such very different sources of radio emissions? The greater distance to Saturn might possibly be one reason, and the fact that Saturn's ring system inhibits radio emissions may be another. But these are long shot guesses: the discrepancy in the strength of the emissions cannot be explained so easily. There is something occurring on Jupiter that is clearly not happening on Saturn.

Uranus and Neptune, orbiting beyond Saturn, are very strong sources of radio signals. Much, however, is still to be learned about each of these planets. Voyager 2 has brought back a wealth of data about Uranus but much of it is still to be cataloged and analyzed. And Neptune is still to be probed. In 1988, we should see a wealth of data published about the planet, as it is the next and last stop on Voyager's tour of the planets.

Uranus was discovered by William Herschel in 1781, and it was such a discovery that is has overshadowed all of his other great contributions to astronomy. Since Herschel, most of the information gathered about the planet has relied on the ingenious theories astronomers and astrophysicists could come up with based on scant observational data, invisible radiation evidence, and a comparative knowledge of the solar system — that is, until the Voyager 2 probe transmitted encyclopedias of information back to Earth about the planet. But even Voyager was not able to develop a complete picture of this giant, gaseous world. Another spacecraft will have to visit again.

For the Neptune encounter, the world's largest radio telescope, called the Very Large Array, is receiving new and stronger receivers to handle the very weak signals coming from distant Neptune. The VLA needs to have a greater signal-collecting area in order to allow for more numerous and more qualitative pictures of the planet. Dr. Paul Vanden Bout, director of the National Radio Astronomy Observatory was quoted in Astronomy magazine as saying that the equipment update on the VLA ". . . not only will enhance the science we get from Neptune, but it will also increase our radio astronomy capabilities after the flyby is over."

The VLA, located near Socorro, New Mexico, is a part of the National Radio Astronomy Observatory. When it first became fully operational in 1981, it was able to synthesize an aperture close to 27 kilometers in diameter and has been extremely useful in producing high resolution radio maps in limited time. The VLA utilizes a total of 27 dishes arranged in Y-formation around a central control room. Each of the dishes is 25 meters across.

Bout and the rest of the astronomical community are very eager about the Neptune encounter because Voyager will be coming about 25 times closer to Neptune than it did to Uranus, so despite Neptune's greater distance, the chances of quality pictures are high. If things go as planned, Voyager 2 will encounter Neptune on August 24, 1989. Perhaps new satellites will be discovered; perhaps, even another planet.

Pluto is well-beyond reach for the present. Discovered by Clyde Tombaugh at the Lowell Observatory in 1930, most of the data that will be compiled about it will depend on electromagnetic analysis. For example, astronomers from the Jet Propulsion Laboratory (Edward F. Tedesco, Glenn J. Veeder and R. Scott Dunbar) and from the University of Arizona (Larry A. Lebofsky) recently completed their own special analysis of Infrared Astronomical Satellite data and have suggested that the planet is 2,200 kilometers in diameter and that its high infrared emissions indicate it is a relatively warm body. This lends further support to the argument that Pluto is indeed a planet and not some wayward moon or asteroid. But the proof is not conclusive. That is, whether or not Pluto is indeed a planet is a question that will not be answered with absolute certainty for some time.

Chances are that in the next few decades, other solar systems will be directly or indirectly observed either as a result of NASA probes or as a result of Earth-based observations. At this writing, astronomers are zeroing on HL Tauri and Beta Pictoris as stars with developing systems. The astronomers are excited about what appears to be disk-like areas of matter orbiting each of these stars. Because both HL Tauri and Beta Pictoris are relatively young stars, the chances of their having advanced planetary systems is highly unlikely. Beta Pictoris has probably only been around for about a billion years and HL Tauri for only about one million. By comparison, our solar system may be 5 billion or more years old. But Steven Beckwith of Cornell University and Anneila Sargent of the California Institute of Technology believe that these stellar environments represent conditions that prevailed when our own Sun was a young and fertile star.

Fig. 5-6. Comets orbit the Sun just as planets do, but their orbits are highly eccentric. This photo is of the comet Ikeya Seki, discovered by two amateur astronomers in 1965. It has a retrograde orbit with an eccentricity of 0.99915 (Courtesy of Lick Observatory).

## COMETS

Comets orbit the Sun very much as do planets and asteroids, but they have highly eccentric orbits. Their sizes can vary considerably, but generally speaking they are only about a kilometer in diameter on the average. Amateur astronomers like to watch the skies in hopes of discovering comets and every year or so amateurs may discover a few of the dozen or so that may be detected. Their great incentive, beyond the thrill of the search, is the fact that comets usually bear the name of their discoverers. (One great exception to this is Halley's comet. Halley did not really discover the comet, but he calculated the orbit and predicted the return of the comet.)

Comets have not evolved very much from the state they were in five billion or so years ago when the solar system was first formed. They are solid bodies consisting of a great deal of ice. They have three major parts: a nucleus from which cometary gas and dust originate; a coma, which is a gas and dust cloud encircling the nucleus; and tails, long wakes of light that can extend for a half-billion kilometers or more and which may consist of dust or plasma (Fig. 5-7).

The great mystery about comets is related to their origin. Astronomers are not sure where they come from. They could be sections of the Sun blown out into space billions of years ago, or just replicas of the birth of our solar system, as described in Chapter 3. Comet Halley's recent appearance — it

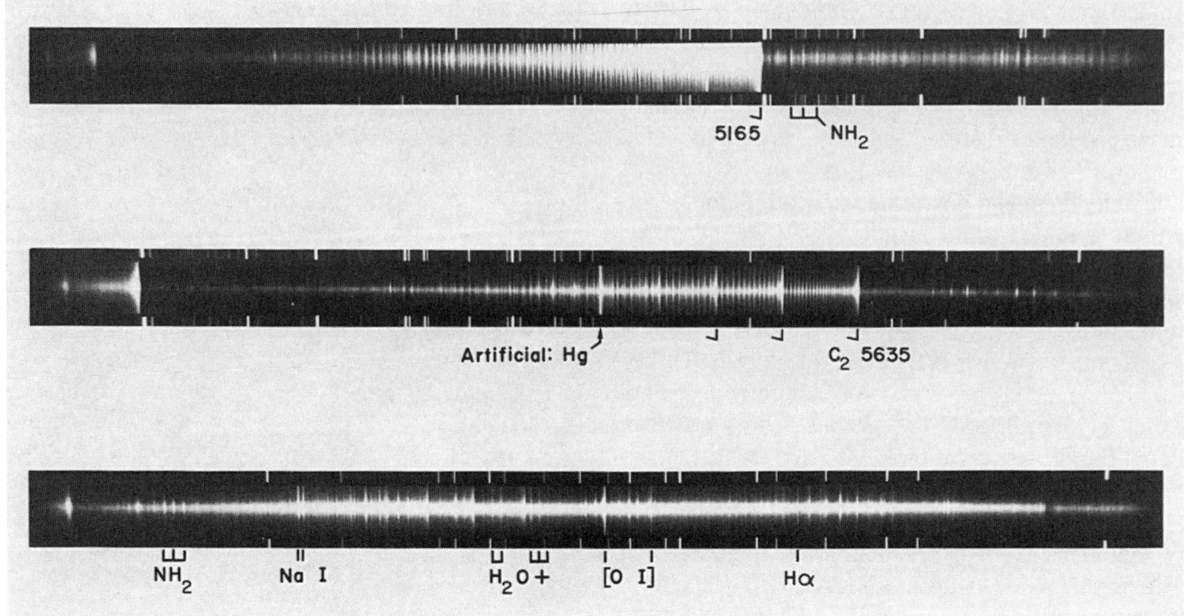


Fig. 5-7. This is a high-dispersion spectrogram of the Comet Kohoutek. It was discovered in 1973 and this spectrogram was taken in January of the following year. It was only barely visible without the aid of a telescope (Courtesy of Lick Observatory).

visits the sky above Earth about every 76 years — has provided special opportunities to add greatly to our knowledge about comets. This last visit, Europe's Giotto space probe gave chase, took pictures, measured the mass, analyzed the coma and the dust tail, and collected data on the interactions between the cometary plasma and the solar wind.[3] When all the data is fully analyzed, there may be surprising news about both the physics and history of comets.

Orbits of comets are so eccentric that they may possibly swing beyond Pluto as certainly as they swing very close to the Sun. When they approach the Sun, the nuclei of comets begin to evaporate under the intense heat, creating a more luminous coma, which, as the comet continues toward the Sun, is blown back over the comet to form the tail. The gases produced by the resulting evaporation are always in a highly excited state and as they begin to settle, they send off radio emissions which can be detected by radio spectrometers and assigned to radio astronomers for analysis.

## THE GALAXY

Whenever we refer to the Galaxy, we are referring to the Milky Way Galaxy, the star system to which the Sun and its solar system belong. The entire population of the Milky Way Galaxy may be three or four billion stars, maybe even more; astronomers vary considerably in their estimates. Many of these other stars may have their own system of planets. When Thales or Aristotle looked up into the sky, every star that they could see was a part of the Galaxy. Things haven't changed. When you and I look up into the sky, each and every star that we see is a member of our Galaxy. Oh, yes, there are billions of stars up there that belong to other galaxies, but we cannot see them; we cannot even see the galaxies to which they belong without the aid of powerful telescopes. And still, with these telescopes there are billions of galaxies that still cannot be detected, for they are far too far away for their light to be collected.

The term Milky Way is also used to refer to the

large and luminous band of starlight that runs around the celestial sphere and is the result of the light from countless and very faint stars. When the term Milky Way is used by itself, it generally refers to this milky celestial view. When it is used with the word Galaxy, it refers to our Galaxy, the one which the Sun orbits the center of about every 230 million years. Everything in the sky is in some orbit. The Galaxy itself may be orbiting some object or system, for all that we know, but the universe is too vast a system for any guesses about just what this or other galaxies may be orbiting. You see, our entire Galaxy is hurtling through space, along with the other galaxies that make up the Local Group. It is moving rather rapidly, too; if current estimates are anywhere near correct, roughly 200 kilometers per second.

By the Local Group, reference is being made to a cluster of galaxies consisting of a few dozen members. The Milky Way Galaxy is only an average-sized member. The more massive members are the Andromeda galaxy and the Maffei 1 galaxy. (Maffei 1 is the brighter of a couplet of galaxies positioned in Cassiopeia near the galactic equator. Andromeda is located in the constellation by that name and may be seen without a telescope, though it appears very, very faint.) Most of the galaxies in the Local Group are dwarf ellipticals or irregular galaxies. Recall that a galaxy is a system of stars, perhaps a billion to 10 billion stars making up that system. So, when we refer to a dwarf galaxy, we are still referring to something that may be 300 parsecs or more. In astronomy, the term "dwarf" may or may not have any relationship to actual or relative size. A dwarf galaxy is basically a faint galaxy. That faintness may be due to low surface brightness or small size, or to conditions. But the term dwarf primarily refers to dimness, so it is possible that an extremely massive galaxy is classified as a dwarf because of its low surface light.

Fig. 5-8. Galaxies are star systems containing billions of stars. This is an edge-on spiral, NGC 4565. Spiral galaxies like this one contain a great deal of gas and dust and are very fertile galaxies, which contain the conditions that allow continuous starbirth (Courtesy of Lick Observatory).

To understand the tremendous size of the Milky Way Galaxy, you must be able to think in numbers that would ordinarily leave you cross-eyed. You know what a light year is; it is the distance that light will travel through a vacuum in one year. If you put that in miles, it is 5,878,000,000,000 miles—not exactly the distance you would like to drive to your summer home. When referencing such fantastic distances as that represented by the light year, it is easy to understand the advantages of using special units of measure, like parsecs, to describe astronomical distances within or without our Galaxy. The parsec is equal to 19.2 trillion miles (30.9 trillion kilometers). This is the distance light will travel in 3.26 years. Suffice it to say that you will never travel one parsec. You cannot go fast enough, and you cannot live long enough.

The Milky Way Galaxy is a spiral galaxy. Spiral galaxies appear as flat disks having spiral arms consisting of interstellar matter and also of stars in the earliest stages of evolution. Astronomers cannot see all of our Galaxy because we are neither at the end of it or located near the center of it. In fact, our solar system may be from 13-17 kiloparsecs from the galactic center. Kiloparsecs—that's a thousand parsecs! This means the Sun may be around 50,000 light years from the center of the Galaxy.

Despite problems of distance and position, astronomers can determine much about the general characteristics of galaxies. For instance, because the Galaxy has a disk-like section in which gas and new stars may be found, there is strong evidence that it is a spiral galaxy.[4] Nonetheless, while a great deal of what we read about the Galaxy, from the number of stars to the great distances between them to their very birth and possible death, is presented as hard fact, much is subject to revision at the next great scientific insight or observation, or after the next space probe.

Whether or not astronomers are tens of parsecs off in determining the size of the Galaxy, the measures with which they are dealing are impressive by any standard. The universe consists of at least a billion more galaxies the same size as ours, and each of these galaxies contains about as many stars. It will be impossible to map the entire universe, for much of it is invisible and a lot of it must be left unseen or undetected because of the relative motions of celestial objects. You may argue this point and say that a detailed mapping of the universe is possible within a century or two, given the great strides taken in astronomy in only this century. But think of this: much of the Earth still goes unmapped — mountains, valleys, forests, oceans, and seas. The universe is far more vast. Astronomers are still trying to map the Galaxy, and after Voyager 2's probe of Neptune, they may have to rework the maps they already do have of the solar system.

As for the Galaxy, estimates are that from the galactic center to the nuclear bulge, the distance may be from 4 to 7 kiloparsecs. The thickness of the nuclear bulge is possibly 2 to 4 kiloparsecs. A halo extends around the bulge; this halo is actually a part of the Galaxy's corona which extends, perhaps, for 100 kiloparsecs. The halo itself has a radius of about 20 to 25 kiloparsecs.

The Galaxy not only consists of populations of stars but also of star clusters. These star clusters are what might be referred to as star subsystems (considering that a galaxy is a star system). They are groups of stars that move together around the Galaxy because they are tied in some gravitational knot. Most of the oldest stars in the Galaxy form the great globular clusters distributed about the Galaxy; these globular clusters are compact, highly symmetrical star groups that probably evolved from the very earliest condensations of the protogalactic cloud that gave birth to the Galaxy. Compared to the stars that make up these globular clusters, our Sun is an extremely early arrival in the Galaxy.

As is the case for most spiral galaxies, most of the radio emissions for the Milky Way Galaxy come from a small region in the nucleus. There are also great clouds of dust and gas throughout the Galaxy, and these serve as relatively strong emission sources. These clouds may extend for distances that can only be measured in light years and are suspected of being the birthplace of stars. As a condensing cloud enters the protostar stage, its core begins to send out a full range of electromagnetic signals, not the least of which are infrared and radio waves. Radio astronomers and their equipment are on the ready to detect the radio emissions and thereby begin to monitor the formation of new stars.

## OTHER GALAXIES

Beyond the Milky Way Galaxy are countless other galaxies consisting of stars, perhaps planets, and a great deal of interstellar gas and dust. The largest galaxies may possibly have diameters ranging in size from 15 to 20 parsecs. All of these great star systems are in constant motion just as is the Milky Way. Not only are their stars and star systems constantly in motion in, around and about the galaxies, but stars are constantly coming into existence or exploding out of existence.

Galaxies are generally grouped according to the way in which they are formed (Table 5-4). There are a number of schemes for classifying galaxies but the most popular seems to be the one devised by Edwin Hubble, whose books on galaxies, stellar systems, nebulae and cosmology still have value.[5] Hubble divided galaxies into four main groups: spirals, ellipticals, lenticulars, and irregulars. Spirals he subdivided into barred and normal.

But galaxies are often given other classifications because of their special characteristics. Seyfert galaxies and N galaxies are just two examples of such additional classifications.

The Seyferts are galaxies with exceptionally bright nuclei and are named for Carl Seyfert, who first classified them in 1943. Seyferts are mainly infrared galaxies, but sometimes they can be the sources of strong radio signals. They represent a phase in the evolution of a galaxy rather than actually a type of galaxy. A lot of the transmissions from

**Table 5-4. Types of Galaxies.**

| Shape | Description |
|---|---|
| Normal Spiral | Flat and disc-shaped star system with fertile spiral arms containing much interstellar matter. These spiral arms are usually the birthplace of stars. The arms appear to come right from the galaxy's nucleus. |
| Barred Spirals | Flat and disc-shaped but here the arms do not begin at the nucleus but rather from a bar (thus, *barred* spiral) cutting across the nucleus. |
| Irregular | No special shape but always relatively small and usually containing large amounts of interstellar matter. |
| Elliptical | Spherical star system having no definite internal structure. These are not very fertile galaxies, as there is little interstellar material. Ellipticals, therefore, are populated mainly by older stars. |
| SO Galaxy | Seem to be a cross between elliptical and spiral galaxies. |

their nuclei are nonthermal. Type I Seyferts have spectra which resembles that of a quasar, but Seyferts are nowhere near as powerful as quasars. Type II Seyferts, like NGC 4151, are particularly strong ultraviolet emitters as well as infrared sources. In the case of Seyfert II's, however, astronomers suggest that most of the activity coming from toward the center of the galaxies is the result of the presence of a black hole.

N galaxies are exceptionally bright galaxies. In terms of luminosity, they fall somewhere between Seyferts and quasars. Unlike Seyferts, many of

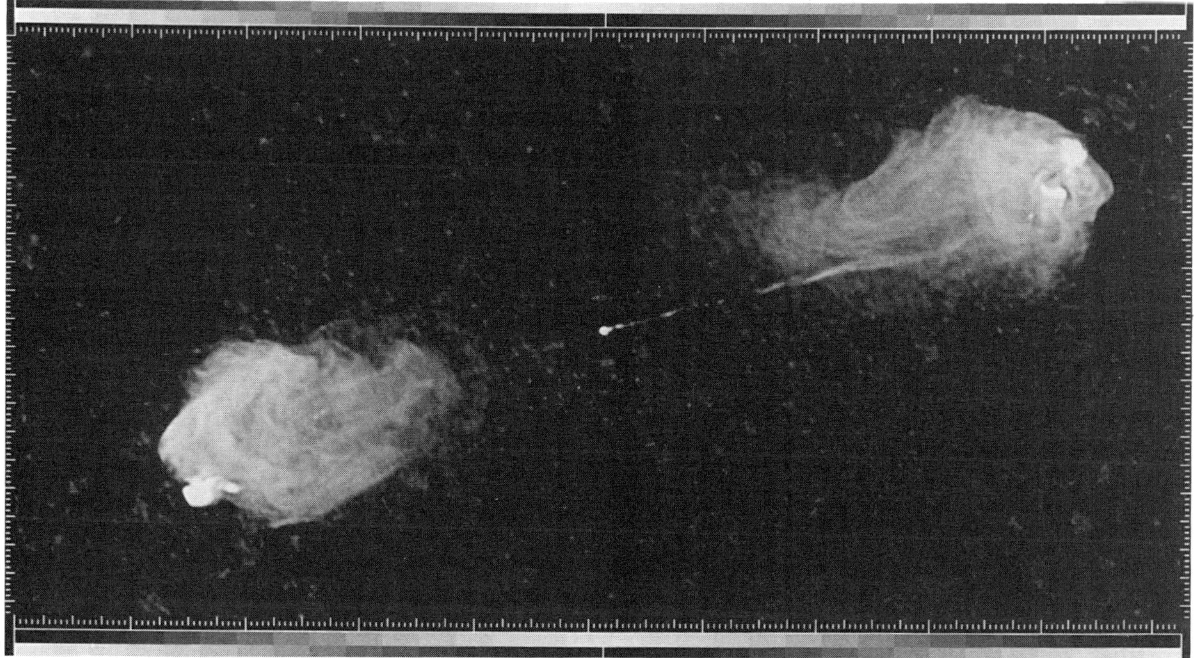

Fig. 5-9. This is a radiograph of the extremely strong extragalactic source, Cygnus A. This image has been processed to emphasize the two radio lobes on opposite sides of the galaxy. (Courtesy of The National Radio Astronomy Observatory, operated by Associated Universities, Inc., under contract with the National Science Foundation. Observers were Richard A. Perley, John W. Dreher, and John J. Cohan).

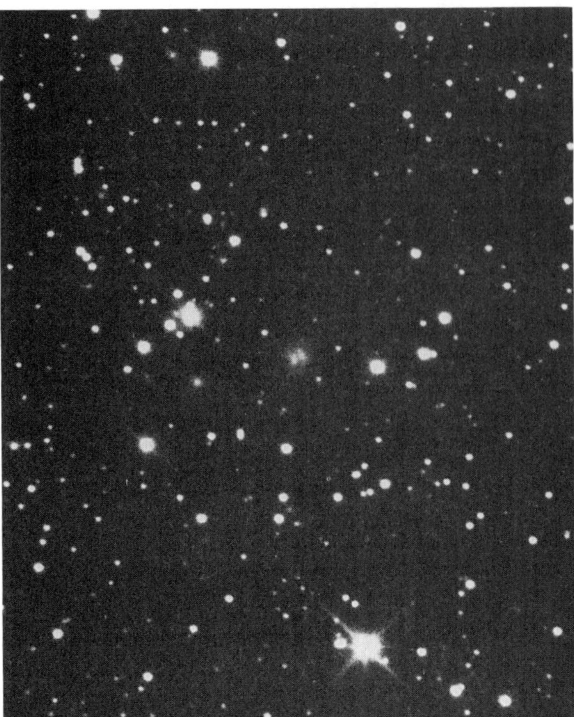

Fig. 5-10. Another view of the radio galaxy, Cygnus A (Courtesy of Lick Observatory).

these N galaxies are extremely powerful radio sources, though there are also radio-quiet N's. These may actually represent an evolutionary stage for a galaxy rather than an actual type.

As for types of galaxies according to the Hubble classification scheme, the spiral galaxies generally appear as flat disks having spiral arms made up of interstellar dust and gas, as well as stars in their earliest stages of evolution. Barred and normal spirals differ only in that the barred spiral has a luminous bar across its nucleus and only the normal spiral actually has its arms starting right at the nucleus. Barred spiral arms seem to come from some line of interstellar matter running from one end of the nucleus to the other. The spirals are often very strong sources of radio emissions.

Elliptical galaxies have elliptical shapes. They are rather dense star systems with many stars well past their prime, stars much older than our Sun. Most of their light is radiated from their centers. The poles of the ellipticals are very flat. Centaurus A, located in the southern constellation, Centaurus, is an example of an elliptical. Centaurus A, roughly 100 kiloparsecs in diameter, is an exceptional strong source of radio emissions, as well as X-ray.

Lenticulars (SO galaxies) are much like the ellipticals except that they have somewhat of an infertile disk. This means that the disk contains relatively little gas and dust. As gas and dust appear to be the necessary ingredients to form a star, these galaxies are not very productive. There are certainly young stars to be found in the disk for the galaxies are not completely barren, but relative to the spirals, young stars are relatively few in the lenticulars.

Irregular galaxies come, as their name suggests, in a variety of shapes. However, they are never spiral or elliptical and they are generally relatively small galaxies. Irregular galaxies often appear to be connected in some unexplainable way, for they all often show similar types of disturbances. These disturbances may possibly be caused by the existence of black holes or else by unusual interstellar actions and reactions. Many of these irregular galaxies literally seem to be on fire.

Exactly how the galaxies have come to take on their different shapes is open for debate. There are at least three current theories explaining how they were formed: the density wave theory, the stochastic star formation theory, and the galaxy-galaxy interaction theory. The density wave theory explains that the spiral galaxies are the result of the instability caused by gravitational interaction of intra-galactic star systems; as the star systems move, orbit and interact, they compress the gases at the disk, thereby causing the collapse of some clouds; these collapsing clouds give birth to new stars. The stochastic theory explains that certain star formations can trigger others through, perhaps, supernovae explosions or the expansion of HII areas. The galaxy-galaxy interaction theory explains that the way in which galaxies interact may possibly cause the more marked spiral designs.[6]

The ellipticals may be the result of collisions between certain components of spiral galaxies. These collisions are believed to greatly alter the form of the spirals. In each spiral galaxy, the collisions result in the destruction of many stars, in changing the patterns of other stars, and in the tearing free of some stars from a spiral's gravitational field. Whatever actually happens, the result is cosmic surgery of a sort that leaves the spirals as ellipticals. Meanwhile, the stars that have been thrown out of the galaxies later catch each other in their own gravitational fields and the result is the formation of irregular galaxies. This is probably why the irregulars are so small compared to the spirals land ellipticals. They are made up of the few stars that escape intact from the colliding spirals.

Because the spiral galaxies contain so much gas and dust and are relatively fertile, they are very strong sources of radio signals, but because of their tremendous distances, the emissions are more often than not rather weak. The elliptical galaxies which consist of populations of old stars, and which are themselves among the oldest galaxies in the known part of the universe, are rarely significant sources of radio emissions unless they belong to that class of galaxies generally defined as radio galaxies. These radio galaxies give off very strong emissions which come from any number of fields within the galaxies themselves. The reason certain ellipticals are more powerful than others is presently unknown, but astronomers speculate that the cause may be related to supernova explosions or the existence of black holes. The ellipticals contain such great populations of older stars that star death can be a very frequent occurrence, even more frequent than in the irregulars which consist of runaway populations of colliding spirals, and certainly more frequent than in the spirals which are made up of mainly new and intermediate aged stars. As most galaxies are suspected of having a black hole somewhere at their center, many astrophysicists like to point to the existence of these holes as the reason for the strong emissions coming from the active ellipticals. As they see it, the black hole, which is strong enough to suck even stars into it, draws stellar gases into a whirling stream about it. This stream of gas traps high-speed electrons that eventually break out of the environment and race away from the center at speeds nearing that of light. These speeding electrons produce something which is called synchrotron radiation. (This is a type of radiation that has many applications in experimental physics, particularly because of its bandwidth, frequency, and natural collimation. It is a form of electromagnetic radiation and is, therefore, a type of energy resulting from the acceleration of electrically charged bodies.)

As the synchrotronic emissions continue their escape and begin to pass through the gas and dust in the immediate environment, they begin to affect emission lines and infrared radiation. But the production of the infrared radiation depends a lot on the density of the gas and dust in the fields just beyond the vicinity of the black hole. If the gas and dust are very dense, the infrared radiation is inhibited. This

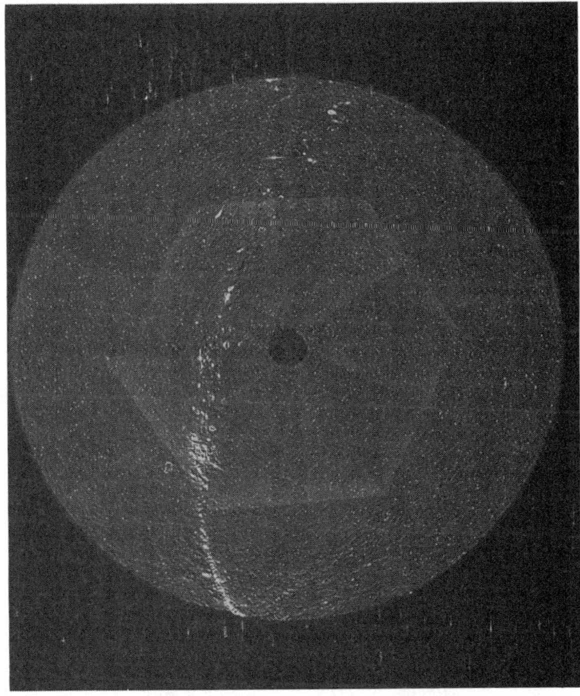

Fig. 5-11. The 300-foot radio telescope in Green Bank, West Virginia took this survey of the sky at 1400 MHz. The survey was of the area between −5° and +82° declination with 12 arcmin resolution. The bright curved band traces the galactic plane, and the faint ripples are diffuse galactic emissions. A close look reveals the existence of ring-shaped supernova remnants. The small stars that appear to be scattered throughout the sky are actually luminous radio sources in elliptical galaxies and quasars billions of parsecs away (Courtesy of The National Radio Astronomy Observatory, operated by Associated Universities, Inc., under contract with the National Science Foundation. Observers were J.J. Condon, and J.J. Broderick).

is possibly why some ellipticals may be highly active radio galaxies and others are relative low or else non-sources: the clouds of gas and dust near the galactic center in each galaxy vary considerably in terms of density. Therefore, the stream of electrons producing the synchrotron radiation may never be able to make it further out into the galaxy where they can interact in such a way with other materials that they produce strong radio emissions.[7]

Actually, beyond the facts that some galaxies are stronger radio sources than others, and that the synchrotronic radiation produced near the core is probably the cause, little more is understood about the reasons for galactic radio emissions.

## QUASARS

Quasars remain one of the greatest mysteries in astronomy. No one is sure about exactly what they may be. Their name is a contraction of the term "quasi-stellar" object. To the optical astronomer, they appear just like any other star in the heavens, yet they seem to be moving much too fast to be stars. To the radio astronomer, both their spectral nature and radio emissions define them as galaxies.

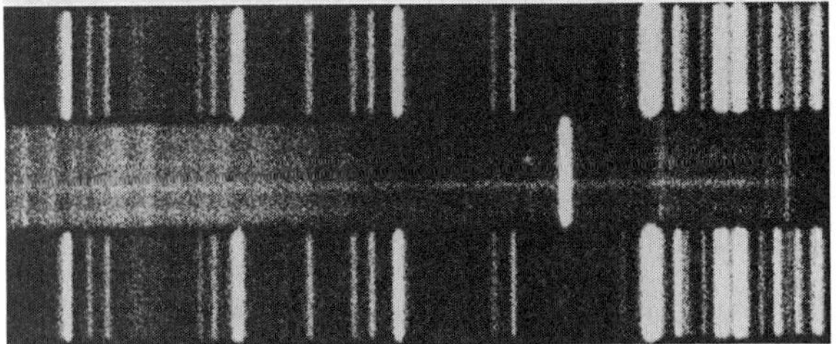

Fig. 5-12. A quasar in the large constellation, Bootes, which may be observed in the northern hemisphere near Ursa Major.

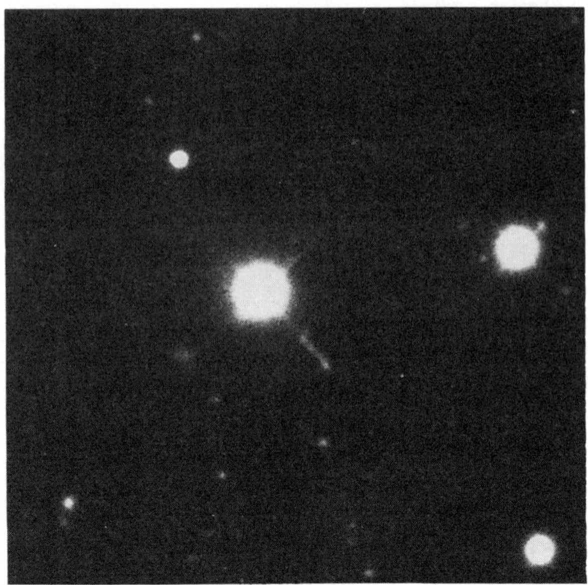

Fig. 5-13. Quasar 3C 273 is one of the strongest radio sources in the constellation Virgo. The quasar is estimated to radiate a hundred times more strongly than an ordinary galaxy, and its jet may possibly be 150,000 light years in length (Courtesy of National Optical Astronomy Observatories).

To say that quasars are moving across or into the sky at very fast speeds is an understatement. Actually, if red shifts are any indication, quasars are moving at speeds near that of light. That is moving! There is a theory in astronomy which states that the distance of objects outside the Galaxy is roughly proportionate to the speed at which they move. This means that quasars are not only the fastest objects in the sky but also the most distant. Being so far away, and yet the source of such powerful radio signals, they must be considerably strange worlds or world-systems far more energetic than even the strongest radio galaxies; they are far brighter than objects many light years closer, and they are far more intensive than many infrared sources.

For years it had been suspected that radio emissions coming from such great distances that the quasars represent could only be the result of supernova explosions, or else exploding galaxies, for what else could possibly exist up there and still signal so strongly? But Scientists were once again overestimating their knowledge of the physical order of things, reluctant to admit that there could be something very different, swifter, and much more powerful than the things they already knew to exist. That is, they were reluctant to admit this until 1961, at which time radio astronomers employed two novel methods to identify these strange quasi-stellar objects, one based on a technique called interferometry and the other based on lunar motions.

There are two basic types of telescopes used in radio astronomy. Both are designed to record and measure cosmic radio signals. They are not too dissimilar, each consisting of systems of dish or linear dipole antennas bringing signals to a peripheral designed as a recording device. When the antenna system consists of independent units relaying signals to one single receiver, then what astronomers have at hand is something very quite like an interferometer. An interferometer is an instrument designed to divide a beam of radiation and send the beams off in different directions, but send them off in such a way that they will transverse and eventually produce interference fringes. These fringes are indicative of radio signals sequencing into the antenna system at different intervals. Phase analyses indicate the true direction of the source to an impressive degree of accuracy.

The two methods of interferometry used are basic and very long baseline (VLBI). In the first technique, a number of antennas are directed at the location of the source. The antennas sense and amplify the signals as they come in and then forward them over physical lines or by microwave transmission so that they can again be united as one radiant beam. In the second technique, VLBI, signals are magnetically taped and then played back in synchronization so that they are produced as a single beam.

The lunar method allows the Moon to run interference between Earth and the radio source. Every time the Moon passes between Earth and a radio source, transmission is blocked. Radio astronomers look for the coincidence between the moment of interference and the coordinates which identify the edge of the Moon at the same exact moment. Later, when the Moon moves out of the line of interference, the signal reappears and the lunar edges are again compared. A comparison of the two different time and space measurements allow extremely accurate determination of the direction to the source.

While quasars have been observed extensively since 1961, there is still not a sufficient explanation of just what they can possibly be. This is mainly because physicists have not been able to find the atomic or nuclear interactions that can possibly propel particles at speeds near those of light. Some scientists speculate that quasars may actually be the nuclei of active galaxies, because many of the quasars estimated to be closer to Earth than others, are surrounded by a haze which may actually be the dimmer and less condensed outer limits of galaxies. It has been particularly difficult for astronomers to come up with any idea of the components of the haze, but there are some indications of a spiral-type of formation, supporting the theory that the clouds are actually the luminous outskirts of the galaxies which surround the quasar.

## SUPERNOVAE

Supernovae are stars which explode with such tremendous force that they not only form an expanding gas shell but also blow their cores, so to speak, into neutron stars or black holes. Neutron stars are stars that have been forced into such a state of gravitational collapse that most of their material has been reduced to neutrons. Black holes are the theoretical end of an object which has gravitationally collapsed. As theory has it, nothing can escape beyond the gravitational field of a black hole unless it is able to move faster than the speed of light. Once any form of matter or antimatter is captured by a black hole, the only thing left to be detected is the mass, charge, and angular momentum of the matter or antimatter.

The expanding gas shell that results from the supernovae explosion is thought to consist of an intense magnetic field and ionized gas. The field and the gas interact to generate powerful radio emissions through the synchrotronic process.

There is evidence of supernovae activity in the spiral arms of galaxies. Spiral galaxy supernovae show a great range of magnitudes. Supernovae only occurring in the spiral arms of spiral galaxies are generally classified as Type II supernovae. They show markedly differing light curves and absolute magnitudes, though most of the Type II's have a maximum magnitude of about minus 17. Spectral analyses show these Type II's to have the usual range of stellar elements.

Supernovae which occur in both spiral and elliptical galaxies, and which have remarkable coincident light curves and magnitudes, are generally classified as Type I galaxies. Spectral analyses for this type of supernova show that the clouds produced from the death explosion contain little or no hydrogen. The gas being ejected into space is roughly equal to one solar mass and is ejected at a speed of around 11,000 km $s^{-1}$; the kinetic energy level is roughly estimated at $10^{44}$ joules.[7]

Astronomers are somewhat more sure of the causes of Type II supernovae. They are believed to have been supergiants of eight or more solar masses and diameters of possibly 10 AU's (2,062, 650 parsecs). These would have been the types of stars that could explode with the kind of force required to turn their massive systems into neutron objects—and produce the radio clouds that signal all the way to Earth.

Astronomers had the opportunity to observe a supernova explosion in the late winter of 1987. Light from an exploding star that met its demise perhaps 50,000 years ago finally reached the planet Earth. The explosion came from inside the Greater Magellanic Cloud, one of our neighboring galaxies. Dr. Robert Williams, director of the Cerro Tololo Inter-American Observatory, reported that spectral data indicated this was a Type I supernova resulting from interactions between a White Dwarf and a normal star.

The exploding star sent strong pulses of neutrinos toward Earth, confirming theories that stars must necessarily produce large amounts of neutrinos, given the particular way in which they are believed to produce fuel through nuclear activity. Collapsing stars on their way to supernova excitement are expected to send off great bursts of neutrino particles, particles which have neither mass nor electric charge and which, strangely enough, occupy no volume of space. This was the first time in centuries that a supernova exploded close enough to Earth to leave a detectable neutrino print (Fig. 5-14).

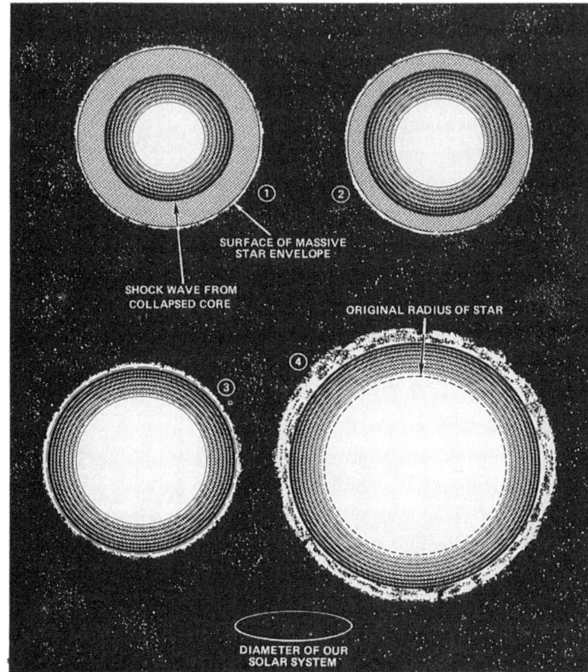


Fig. 5-14. In this illustration from the National Optical Astronomy Observatories, we witness the death of a giant star by way of supernova explosion. This illustration came about as astronomers from the United States, Chile, and Argentina observed a massive star in galaxy NGC 4699 exploding. This particular type of supernova begins as a very massive star, probably a red supergiant a thousand times larger than the Sun or possibly a Wolf-Rayet star, which are very luminous, very hot stars with temperatures about 50,000 degrees Kelvin. The sequence in the illustration may be described as follows. (1) The star is detected June 6, 1983. Its core has collapsed and the resulting internal explosion has sent a shock wave over 5 billion miles to the surface. (2) Spectra is obtained at Cerro Tololo on June 14. The shock wave is now 80 percent of the way to the surface. (3) The supernova reaches maximum brightness; this is the time when supernovae are usually detected. The shock wave is almost at the surface and there is hydrogen in the thin stellar shell soon to become part of the advancing explosion. (4) The star flies apart and by late July it has a radius of close to 8 billion miles.

## IONIZED HYDROGEN CLOUDS

Still other sources of strong radio emissions are the large clouds of interstellar gas that consist mainly of ionized hydrogen. The ionization is expected to be the result of the emissions from the very bright and very hot stars that the clouds cover. These clouds send very strong radiation at the visual wavelengths, but they are also very powerful sources of emissions at radio wavelengths.

The ionized hydrogen clouds are referred to as HII regions. The clouds extend for fantastic distances but are not usually more than 200 parsecs in diameter. They are one of the means by which astronomers determine intergalactic distances.

Just as the radiation from these massive ionized hydrogen clouds can be detected, so can radiation from individual hydrogen atoms. This type of radiation is the product of proton and electron interactions, in which the same particles are suddenly forced to assume identically directed spins by some action of the atoms they inhabit. They produce emissions that can be detected at wavelengths of 21.1 cm. The possibility of detecting radiation from individual hydrogen atoms was predicted in 1944; seven years later actual detection was accomplished. Hydrogen line radiation has become an important research area in astronomy because of its cosmological impact. Given that cosmology is fundamentally the search for the order (including the origin) of the universe, the more astronomers can piece together the universal puzzle, the better their understanding of the cosmos. And as the universe is fundamentally a system of galaxies, the more that can be learned about galaxies the better.

Because a lot of what scientists can determine about other galaxies is dependent a great deal upon what they can learn about our Galaxy (just as what they learn about the stars depends a great deal on what they can find out about our Sun), the better the understanding of the Milky Way Galaxy, the better the understanding of star systems in general. And astronomers have been able to learn a great deal about our Galaxy by measuring the relative amount, and velocity of, hydrogen in selected galactic areas. Analyses of the radio emissions coming from these fields has helped astronomers verify the Galaxy's shape, identify and compare other galaxies, and estimate what percent of the Galaxy's total mass is made up of interstellar gas (5-10 percent).

There is a great cloud of ionized hydrogen gas floating about 12 billion light years away which may be indication that a new galaxy is beginning to evolve. This new star system has been designated 3C 326.1 and spectral studies indicate that the galaxy so far has a very sparse population of stars. But 3C 326.1 is another one of those cosmic contradictions that astronomers are constantly stumbling across. While radiation studies indicate the galaxy is at a rather early stage in its evolution as a star system, in terms of its actual age, it has been around a long, long time. This may mean that a lot of ideas about galactic development may be blown away. Galaxies are believed to have formed very early in the history of the universe. 3C 326.1 may prove this to be a fallacy.

## MOLECULAR CLOUDS

The more accepted theories of star formation explain that the "parents" of these stars are massive nebulae and some star, group of stars, or other object, that may explode with such force that the mother cloud is forced to contract. These huge clouds which contain the "womb" in which the star will be conceived are great conglomerations of cool and dense interstellar gas and dust. These clouds are so dense that the atomic population becomes molecular. The carbon monoxide in these clouds gives off radio emissions, emissions that remained undetected until the 1970s. Carbon monoxide is much more easily detected than hydrogen; however, there is significantly less carbon monoxide floating in interstellar regions than there is hydrogen. Nonetheless, the emissions are strong enough and in enough quantity to allow astronomers to read the clouds, no small feat as these clouds can extend to 50 parsecs. In addition to carbon monoxide, there are dozens of other molecular structures in these clouds.

There may be as many as 10,000 of these molecular clouds in the Galaxy, all possible birthsights for the many stars yet to be born. Many of these clouds may be "barren," but that does not mean they will always be. Nature has a way of using everything in good time. One day a supergiant, parsecs away, will die in one of those fantastic explosions discussed earlier, and in doing so will cause enough disturbance to start the cloud in its birth cycle. Its death means life for another; perhaps this other will be a star, perhaps it will be an entire galaxy. The cycle of birth and death is as continuous in the heavens as it is here on Earth.

The molecular makeup of these clouds is best examined with a radio telescope that utilizes a line receiver capable of detecting the narrow bandwidth emissions that come from molecular sources of radiation. Until radio astronomers zeroed in on these molecular sources, the discoveries were few and far between. Actually, the first interstellar molecules (CH, CH+ and CN) were first detected by optical means.[8] Now, however, the radio discoveries are frequent. Table 5-5 gives a partial list of these molecules and indicates that most of them, surprising enough, are organic rather than inorganic.

The emissions which radio astronomers detect are usually the result of the rotational characteristics of the molecules. Their wavelengths are limited to the gigahertz (one billion hertz) region of the radio spectrum. Astronomers are expecting that molecular astronomy will help determine the chemical and thermodynamic formulas that result in the "impregnation" of a cloud, the resulting protostar stage, and the eventual "birth" of a star.

**Table 5-5. Some of the Molecules Detected in Space.**

| | |
|---|---|
| Ammonia | Formylium |
| Carbon Monosulfide | Hydrogen Cyanide |
| Carbon Monoxide | Hydrogen Isocyanide |
| Carbon Sulfide | Hydrogen Sulfide |
| Cyanamide | Hydroxl |
| Cyanoocetylene | Isocyanic Acid |
| Cayanodiacetylene | Ketene |
| Cyanoe Thynyl | Methyl Formate |
| Cyanohexatriyne | Methanimine |
| Dimethyl Ether | Methyl Acetylene |
| Diazenylium | Methyldadyne |
| Ethanal | Methyl Alcohol |
| Ethynyl | Methylamine |
| Ethyl Alcohol | Methyl Cyanide |
| Ethyl Cyanide | Sulfur Dioxide |
| Formaldehyde | Thioformylion |
| Formamide | Vinyl Cyamide |
| Formic Acid | Water |

## MOLECULAR ASTRONOMY

There is a branch of astronomy which deals specifically with radio emissions from molecules inhabiting space called molecular-line radio astronomy. It has its origins with the 1968 discovery of hydroxl, ammonia, and water in dark interstellar regions. Since that time, many additional molecules have been discovered. Most of the molecules detected to date are carbon-based and organic. These molecules are in the gigahertz region of the electromagnetic spectrum and are important because they are helping astronomers learn much more about possible ways in which stars can be formed. There is still a great many missing pieces in the puzzles of stellar evolution—and molecular-line astronomers may be the one to find most of these missing pieces.

## PULSARS

Pulsars are sources of highly consistent pulsating radiation (Fig. 5-15). They were first discovered by radio astronomers in 1968, and they've been a conundrum ever since. Most of the studies of pulsars have been conducted through radio astronomy but some pulsars have been detected at visual, gamma ray, and X-ray wavelengths. Pulse periods are highly constant but pulse intensity can vary considerably over extensive periods. Nonetheless, astronomers have come across one pulsar that signals so consistently that we now have something that is called *pulsar time*—in addition to atomic time, universal time, sidereal time, and dynamical time. With most pulsars, however, pulse ranges run the gamut from less than one msec to 20 msecs. At what interval the pulses are detected depends upon their wavelength; that is, pulses at the longer wavelengths generally arrive later than those at the shorter wavelengths.

There have been estimates that the number of pulsars actually existing in the Galaxy may number into the millions. A dozen or so may appear or reappear every one hundred years. They are believed to be stars well into their evolution, or they may be

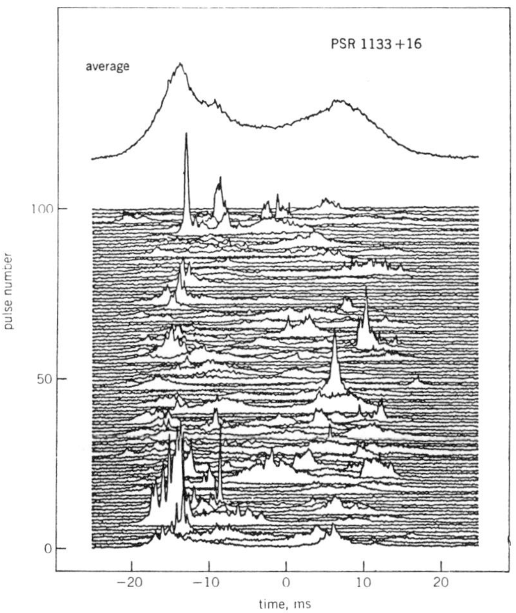


Fig. 5-15. The variations that take place within pulses and from one pulse to the next in signals from pulsars. The source pulsar is PSR 1133+16 (Courtesy of McGraw-Hill Encyclopedia of Astronomy, Sybil B. Parker, Editor, 1982).

the remnants of supernovae. They have burned up their internal structure and depleted their energy sources, and now they are beginning to collapse. These are very dense stars, possibly neutron stars. The signals they emit are the result of synchrontronic emissions. Exactly where in the dying, or neutron, star the signals are originating from is still a subject for speculation. There are many possibilities, the more probable being the magnetic poles.

## MASERS

Masers are highly limited radiation sources emitting at the higher frequencies. Generally, electromagnetic radiation may be detected at frequencies between 3000 hertz and $10^{22}$ hertz. Maser frequencies are rated in gigahertz. Maser is a word derived from the abbreviation of the term *m*icrowave *a*mplification by *s*timulated *e*mission of *r*adiation. The term is rather new in astronomical glossaries. It was first used in the early 1950s but probably did not begin to appear in print until around 1955.

Basically, masers are any type of celestial object that generates electroradiation in the microwave region of the spectrum; this radiation is the result of the natural swinging motion of atoms or molecules. They have been detected in a number of cosmic fields, including molecular clouds and in regions of older variable stars. (Variable stars are those which show relatively high variations in brightness, radial velocity, and spectral characteristics over a period of time.)

Masers are easily detected by radio astronomers because they are relatively high radiators of radio signals. The signals are highly variable and the emitting regions are usually less than one parsec in diameter—it still takes about three light years for a signal to go to extreme points of the maser.

## RADAR ASTRONOMY

Radar astronomy is also a relatively new term in our language, dating back to about 1959. Its field is currently limited to our solar system but by the beginning of the next century (not too far away) it

**Table 5-6. Quasars, Pulsars, Masers.**

| Object | Description |
|---|---|
| Quasar | Celestial object appearing to be a star which cannot be because of its tremendous distance and powerful radiation. |
| Pulsar | Celestial object pulsating radio, X-ray, and optical radiation at relatively short intervals (roughly 1.3 of a second). Pulsars are believed to be neutron stars. |
| Maser | Celestial object emitting at the microwave wavelengths as the result of atomic or molecular level oscillations. |

may be used to explore the Galaxy. So far, radar astronomy's most important contributions have had to do with establishing intra-solar system distances. Thanks to radar astronomy, most of the measures of the planets and their distances are very precise, whereas distances beyond the solar system begin to become more like guesswork (very educated guesswork, however).

Radar astronomy has also been responsible for establishing the currently accepted value for something called an *astronomical unit* (AU). Not long ago, the AU was based on a formula related to the Earth's axis, but now it is defined by Kepler's third law. An AU is 149.6 million kilometers (or 8.3 light minutes). This is the average distance between the Earth and the Sun.

Fig. 5-16. This early Bell Telephone photograph shows the construction of the radio telescope in Holmdel, New Jersey (Courtesy of AIP Niels Bohr Library).

Fig. 5-17. Now that science has moved into the space age, it will take its advanced detecting equipment with it. Actually, it has little choice. While radio astronomy can remain landbased, the other astronomies, like ultraviolet, X-ray, and gamma-ray, must get above the Earth's atmosphere to detect the respective electromagnetic radiation, although this does not hold true for all detection in the ultraviolet ranges of the spectrum (Courtesy of NASA).

# Chapter 6
# Microwave Cosmology

MICROWAVES ARE HIGH FREQUENCY RADIO WAVES that lie between the radar and infrared regions of the spectrum. The microwave wavelengths are from $3 \times 10^{-4}$ to $3 \times 10^{-2}$m, and the frequency range is roughly $10^{10}$ hertz to $10^{12}$ hertz. Thus, microwave detection and study remains in the realm of radio astronomy. Other chapters deal with radiation that is being emitted from celestial objects; this chapter deals with a type of radiation that astrophysicists believe is not being emitted by astronomical objects but rather appears to be associated with cosmological events. Because of this, microwave background radiation may be the means by which cosmologists can travel to the beginning of time.

## COSMOLOGY

Before understanding the importance of microwave background radiation, it is necessary to understand the distinction between astronomy and cosmology, and also something about the Big Bang theory of the origin and evolution of the universe.

Actually, the Big Bang theory is only one of the major theories that have been devised to explain the beginning, growth, and possible extinction of the entire universe, but it is currently the theory that seems to be at the forefront of cosmological arguments. Currently, it is undergoing some serious modification and within the next decade or so it may be simply part of a grander theory. Right now, however, despite its flaws, the Big Bang model of the universe is the one most astronomers and cosmologists favor.

Cosmology is a theoretical science that attempts to develop a comprehensive theory of the why and how of the universe. Astronomy, on the other hand, is a science concerned with accumulating facts about the nature, motion and interactions of all celestial objects. Cosmologists launch their investigations from a pad of astronomical data but they operate in a theoretical theater, as they must, for they are concerned with information that no one will probably ever be able to prove or disprove. The cosmologist is concerned with things that happened at the beginning of time, or occurred millions of years ago, or will occur millions of years from now—in times when there was probably no witness to the events

Table 6-1. Microwave Astronomy/Cosmology.

| Characteristics | Descriptions |
|---|---|
| Definition | Astronomy at high frequency radio frequencies between radar and infrared ranges. |
| Frequency range | Approximately 1000 to 300,000 megahertz. |
| Wavelength range | Approximately 300 to 1 mm. |
| Microwave sources | Masers, primordial cosmic. |

or when there will no longer be anyone around to witness events. But from a purely philosophical perspective—and certainly from a theological one—past existences (and other universes) and the immortality of the human race as well as the eternal nature of the world remain subjects for serious debate.

## THE BIG BANG THEORY

The dominant cosmological theory holds that everything which exists in the universe originated in an explosion that occurred twenty to one hundred billion years ago. At the time of that explosion—no one knows what actually exploded—the universe was at its maximum temperature. Since that time, it has been growing cooler and cooler. What it was that exploded must forever remain a mystery, if there was anything at all. Anything at all? Yes, some theorists hold that in the micromoment before the Big Bang, nothing existed—not energy, not time, not space. The entire universe—all of creation—is argued to have begun at the very moment of the Big Bang. Cosmologists speculate about events right to that point of the bang and leave additional questions about how nothing at all could have exploded to the philosophers and theologians, although many scientists hold that our universe may be the result of a section of another and greater universe that one day separated from the mother system and eventually exploded in a hot ball of fire. Once the explosion occurred, however, items were not simply hurled into space but rather began to move outward because the space between them was stretching. Space to physicists is not some stretch of emptiness, you must remember, but is instead some type of elastic medium that can expand and contract.

In its very early stages—less than a second after the Big Bang—the universe was a masterpiece of simplicity. It was simple in the way that an acorn is simple, but just as that tiny acorn can contain all quantities and qualities necessary to produce a great oak tree, the universe in its most elementary form harbored the energy and the matter that would one day bring about fantastic star systems and life.

Table 6-2. Major Cosmological Theories.

| Name | Argument |
|---|---|
| Big Bang | All of what is included in the known universe has its origins in an explosion that occurred billions of years ago. This was not an explosion in the terms of which we might ordinarily expect in that matter was not hurled out into empty space as a result of it, but rather space itself was caused to expand. |
| Steady State | Space is flat, is the same in all directions, and the universe is the scene of constant creations of new matter at a rate of, perhaps, $10^{-10}$ nucleons per meter cubed per year. |
| Cyclical | The universe is in a constant state of expansion and contraction. At the present time it is expanding. |

Fig. 6-1. Everywhere that astronomers point their telescopes, there are stars. But how did they come to be? How did the universe come to be? These are questions the cosmologist hopes to answer. This photograph is of an open star cluster in the constellation Cancer designated NGC 2682. The cluster is believed to have been around for 75 percent of the life of the universe. It is about 2500 years away, is 12 light-years in diameter and contains approximately 500 stars (Courtesy of National Optical Astronomy Observatories).

Moments later—less than a second after the Big Bang—elementary particles began to appear as the result of the thermal radiation produced. By this time the universe passed into and through an inflationary age of exaggerated expansion, and now had settled into a steady and predictable stage of expansion and evolution.

By the first second, the universe would have grown to about the size of the Earth today and gravitation, electromagnetic, strong nuclear, weak nuclear, and possibly antigravitational forces would have become separated. Five minutes after its birthtime, the universe cooled significantly and contained an abundance of helium. After twenty billion years (or whatever is the current guess), the universe became the system of galaxies and clusters of galaxies we observe today.[1]

The Big Bang theory is a new theory only from a technical standpoint. Actually, the idea of the world (universe) starting suddenly from nothing at all can be traced to Jewish Scriptures and Greek philosophy. The idea that the start occurred with a great explosion, and the development of the scientific theories that have allowed for a step-by-step account

**Table 6-3. Big Bang Cosmology.**

| Time | Activity |
|---|---|
| Zero | A great explosion causes space to expand. No one can possibly be sure of what actually caused the explosion—perhaps it was some primeval atom or perhaps a catastrophe in another universe that preceded ours. |
| $10^{-43}$ | The universe is still extremely hot though it has cooled extensively from Big Bang temperatures. Now, it is smaller than the Earth but it contains the physics for all that can ever exist. |
| 1 second | The universe has grown to the size of the Earth and elementary particles are in existence. The basic forces of nature are now independent. |
| 5 minutes | Helium is in existence; temperatures have cooled drastically. |
| 1 million years | The very beginnings of what someday will be galaxies of stars and planets are now happening. |
| 20th Century | Temperatures in the universe have cooled to roughly 3 degrees Kelvin on the average. There are hundreds of billions of galaxies in existence; there is life on at least one planet; and you and I are here. |

since about time $10^{-43}$ are very new. Among the evidences given to support the idea of the Big Bang are the recession of the galaxies, the discovery of radio galaxies, and cosmic background microwave radiation. This does not mean that there are no strong arguments against the possibility of the Big Bang; these are discussed later.

Applying the Doppler effect, astronomers have been able to determine that the most distant galaxies are receding pell mell from Earth and from each other. This is interpreted as meaning that the universe is currently in a state of expansion. There is a lot of room for challenging this thesis, as there is room for questioning all cosmological models. Galaxies which show evidence of marked redshifts are tens of millions of light years away, so redshift data must be questionable. The distances to these galaxies are just too great to allow dependable redshift readings. And beyond this basic objection is the "tired light" theory originally presented by Fritz Zwicky and brought to the forefront by Paul Violette in an article in the February 15, 1986 Astrophysical Journal. Because light from very distant celestial objects must necessarily suffer an energy loss, Doppler redshift readings would naturally show up regardless of whether or not the galaxies were receding.[2] This certainly makes a great deal of sense, and the argument tells us that not only is it possible that the universe is not in a phase of expansion but that, if it is, this expansion is a lot slower than currently believed.

On the other hand, the research in radio astronomy appears to confirm the idea that the galaxies are receding. Radio astronomy now employs telescopes that are superior to optical instruments in the collection of reliable astronomical data. The radio signals that are being picked up seem to confirm that once upon a time, the universe was far more tightly packed than it is today. This is another way of saying that the galaxies were much closer to each other in the past and so, therefore, the universe must be in a continuing phase of expansion. High energy radiation studies further confirm this idea.

Confronted with the logic of Zwicky's "tired light" argument as well as the evidence from radio astronomy, astronomers would ordinarily be reluctant to lean one way or another toward the Big Bang theory, except for the discovery of microwave background radiation and its implications. As the theory goes, if the universe was born in a great fireball that occurred 20 billion or so years ago, even that length of time is insufficient to erase all traces of the radiation emitted during the event. After all, as physicists estimate it, the temperature of the universe at birthtime was perhaps 200 to 500 billion degrees, possibly even more. The background radiation associated with the event, then, must still be around.

Fig. 6-2. The sky appears to us as a field of tiny lights, each light representing a star or, in some instances, a planet. But as we zero in with telescopic instruments, what at first seems to be only one bright object is really a population of countless stars, as this photograph of a globular cluster clearly illustrates. Everything in the sky may have its origin in some primeval atom that exploded 15 to 20 billion years or more ago (Courtesy of Lick Observatories).

In the early 1960s, two physicists working at Bell Laboratories were utilizing a state-of-the-art radio antenna and receiving system that Bell had at its plant in Holmdel, New Jersey. The physicists, Arno Penzias and Robert Wilson, were attempting to measure radio signals that appeared to be coming from various galactic sources. They wanted to find out the locations of the emissions and possibly their causes. As they searched the sky with their equipment, it became evident to them that the signals they detected were not coming from separate astronomical sources. Instead, the emissions seemed to be coming from every possible point in space. This seemed to be impossible, so Penzias and Wilson continued to monitor the receptions. However, even after months of tracking and measuring the signals, specific sources or locations could not be discovered and the signals did indeed seem to be homogeneous.

Still, doubt persisted, for how could this be so? Perhaps, there was something wrong with the equipment! Troubleshooting efforts proved the equipment was ship-shape and led both Penzias and Wilson to finally accept that what they had detected was the microwave background radiation created during the very moment when time began, when the universe was created in one fantastic explosion. Now, it may seem incredible that scientists would accept that this microwave background radiation can be traced back to the moment of the Big Bang, what with the tremendous populations of celestial objects and all the radiation and other noise that is constantly being generated, but they were, nonetheless, duly impressed by the experiments in Holmdel and Penzias and Wilson were eventually awarded a Nobel Prize in physics. They had discovered what might be considered the very first broadcast.[3]

The case is still not closed; the Big Bang is not fact, but theory. Incorporated in the idea of microwave background radiation being substantial proof of the Big Bang is the theory of a homogeneous universe—at least on an overall scale, if not in localized areas. And recent research has turned up clues that challenge this idea of a homogeneous universe. It appears that galaxies are distributed about the surfaces of massive, bubble-like cosmic voids, at least according to a three-dimensional map constructed by members of the Harvard-Smithsonian Center for Astrophysics.[4] Contributing researchers were John Huchra, Margaret Geller, and Valerie de Lapparent. They expect that the bubbles will act and interact in such a way as to behave much like ordinary soap bubbles. As their model suggests, the enclosed areas of each of these cosmic bubbles are empty and the only place celestial material can be found is at the intersections of bubbles. Their map seems to indicate that the universe is a lot less dense than astronomers have believed and, therefore, the universe is "open" and programmed for infinite expansion.

However, the universe may not actually be expanding in the predictable way that astronomers expect. In the July 1986 edition of *Astronomy Magazine* there is a report by American and British astronomers that certain galaxies appear to be moving out into space in some random course. Now, all galaxies have random motions as a result of gravitational influences of other giant star systems, but the galaxies under study by the British and American teams are clearly moving at faster speeds than should be expected given the accepted expansion rate of the universe. The gravitational action of these streaming galaxies is also rather strange. They seem to be less the source of gravitational interaction than, perhaps, the victim of it. This leads astronomers to guess that there is some massive celestial system pulling the streamers in the same direction of the Milky Way, or that the galaxies are still under the influence of primordial explosions that also caused the bubble-like voids reported by the Harvard-Smithsonian group.

## CYCLICAL THEORIES

On this question of whether or not the universe is open as much as current research indicates, or is closed as some still argue, it is interesting to note that some theorists suggest that we live in an oscillating universe; that is, we live in a universe that goes through cosmic cycles of expansion, contraction, and re-expansion, and this has been going on for eternity. This is a modern version of an old

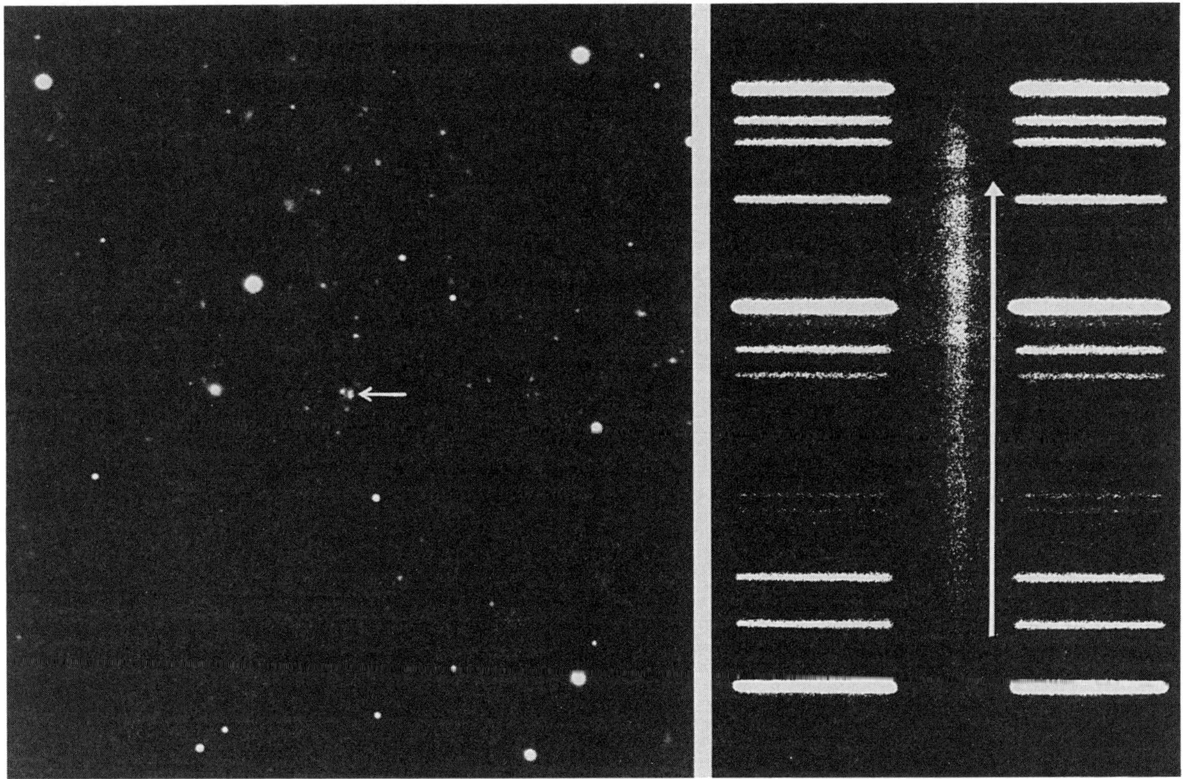

Fig. 6-3. This Hale Observatory photograph is of a hydra cluster of galaxies and indicates the particular galaxy observed for velocity measurement. The accompanying spectrum shows a large redshift equal to about 38,000 miles per second.

philosophical argument for something called "eternal return."

For more than a thousand years, some philosophers have argued that there is indeed an eternal cosmic cycle, which they often refer to as cyclical recurrence, that assures *universal* events will be continuously repeated. Just as the universe was born as the result of some unexplainable cosmic phenomenon, so, too, it will one day return to its beginning state, but only to be reborn by the same process which first brought it into being. With each rebirth, supporters of the theory have argued, there must necessarily follow the same cosmic patterns that would result, just as before, in the creation of the same matter that will evolve into the same galaxies of stars and planets, and into the same organic forms we witness today. The theory holds that the same individual lives that existed in times past will exist again. The numerical order continuing with each rebirth must always repeat itself, and Aristotle, Copernicus, Laplace—even you and I—will all live again. But this does not mean that all our lives must follow the same patterns as before, for we are individual beings with free wills and the ability to effect change, and our lives can be different with each rebirth.[5]

Plotinus supported this idea in his *Fifth Ennead*, in which he writes that "If it is true that Socrates' soul is eternal, then the Authentic Socrates must exist always; this is to say his individual soul has its existence not only in this world but in the Supreme." Plotinus goes on to explain that every individual person inherits something which he calls the *Reason Principle*, which exists in the cosmos, and these principles can never change, whether the universe is in a period of expansion or contraction or at its original state. Scientists and philosophers have always enjoyed toying with this idea of cosmic cyclical recurrence. In science, of course, Plotinus' *Reason Principles* are the physical laws which govern the universe, and given the same set of conditions that must prevail each and every time a new cosmic cycle begins, there must be the same products, just as with any chemical experiment, the results can always be predicted. Men like Newton, Descartes, Spinoza, and Kant, though they acknowledged natural periodic cycles, never saw sufficient grounds for arguing that this cyclical recurrence applies to the entire universe.

Cyclical recurrence does indeed have its place in cosmological theories even today. Cosmologists readily consider the possibility of an oscillating universe, although at present there are no facts to support the theory. If the universe is not open and, therefore, not designed to expand forever, then at some time it must begin to close (or contract). This contraction phase will continue until the universe returns to the very singular existence it was at the start of time—or was at since ever (another way of saying "since eternity.") The same basic laws that once governed its expansion from a very simple and highly compact state must necessarily remain to be applied again and a new fireball should result. Of course, cosmologists stop far short of inferring personal immortality in their cyclical scheme—quite unlike the philosophers—and speak only of a highly possible, though not probable, cosmic recurrence.

## THE STEADY STATE THEORY

In 1948, the idea that the universe was in a steady state—eternal and roughly—the same everywhere was proposed. The theory, however, did not infer that the universe was unchanging, for as the universe expands, new matter must be created to fill in the voids. The theory held great sway until the discovery of microwave background radiation finally put it to rest. Proof of the existence of microwave background radiation has been interpreted as being proof that the universe had an intensely hot beginning and the steady state theory cannot account for the presence of microwave radiation.

In recent years, however, it has begun to become apparent that the background microwave radiation is not quite as steady from all regions of the universe. Thus, there must be a great deal more to learn about its cause and what can affect it; satellite research planned in 1987 or 1988 should once and for all ascertain whether or not the microwave background radiation can be associated with the Big Bang.

# Chapter 7

# Infrared Astronomy

USUALLY BY THE TIME SCIENCE ARRIVES AT SOME new discovery or begins to employ new technology, it finds that Nature is already there with its own methods. For instance, Nature designed creatures for flight and submarine life millions of years before man even considered these possibilities for himself.

It is the same for infrared technology. Physicists have learned how to detect electromagnetic radiation in the 0.8 micrometer to 1000 micrometer range only now to discover that Nature (or, again, the incredible intelligence that lies behind Nature) is already there. Nature seems to have seen fit to give the ability to detect infrared emissions to many of its nocturnal creatures. There is no conclusive proof, of course, that some animals and reptiles can see at night by infrared detection but there is evidence for the idea. Take snakes, for instance. They are clearly equipped to detect the heat given off by nearby animals and, therefore, to track them down even in the absence of all light.

## INFRARED TELESCOPES

The infrared radiation (as well as radio radiation) that comes from celestial objects, is at wavelengths that are greater than the particles which make up space dust. Thus, these emissions are able to make their way to the surface of the Earth. It is this space dust that inhibits most of the other forms of radiation; gamma, X-ray, and ultraviolet radiation is too easily blocked by the dust particles.

The detection and study of infrared radiation from space has become possible because of the great technological advances that have allowed the construction of the telescopes needed in infrared astronomy. These telescopes are often called *flux collectors* and they must be reflecting telescopes; the refracting kind are useless at the infrared wavelengths. Infrared telescopes must be installed at extremely high altitudes because Earth's atmosphere, which is very vaporous and consists of a great deal of carbon dioxide, prevents many of the infrared rays from getting through. Fortunately, there are atmospheric windows which allow astronomers to detect a great deal of infrared radiation from ground points. These atmospheric windows are fields in the electromagnetic spectrum that are not generally absorbed or reflected by Earth's atmosphere.

Infrared radiation was actually discovered back in the nineteenth century. It was probably the work of Isaac Newton that led to its eventual discovery. Chapter 2 explained that Newton arranged an optical system which captured white sunlight in a small slit in a window shade and projected it through a prism. This experiment was conducted in 1666 and it was the basis for additional experiments by Newton and other scholars into the nature and characteristics of light and color.

Well, 134 years after Newton's ingenious scheme, William Herschel used a similar experiment to capture sunlight and spread it out for analyses. Herschel, however, was not at all interested in the refractive characteristics of light. What he was interested in was the dispersion of heat in the solar spectrum. He was after information that would help him in the design of more efficient telescopes. Once he had the solar beams separated, as did Newton almost two centuries before, he placed a thermometer next to each color. He fully expected the thermometer to show varying measures as he moved from color to color in the spectrum, registering high at the red end of the spectrum and low at the opposite side. But to his astonishment, he discovered that the maximum temperature was registered on the other side of the red wavelengths. The highest temperature he was able to measure actually lay outside the visible portion of the spectrum, beyond that of red light. Herschel had provided the very first evidence of infrared rays, a name given to this unseen radiation by Herschel himself. Little could he guess how important infrared detection would be in twentieth century astronomy.

Now the telescopes employed by astronomers for the detection of infrared rays need not be the precision instruments that are required in optical astronomy. This is due to the length of the waves being detected, which, because they are longer than those in visual radiation, are much more easily detected and measured. Still, the infrared wavelengths that are the most easily detected are those at the shorter wavelengths of the infrared range.

Infrared astronomers are faced with a very special problem that optical astronomers need never worry about. This has to do with the photographic end of the science. Photographic plates available today have no sensitivity to the longer wavelengths in infrared, so special instruments must be employed.

One of these instruments is something called a photovoltaic detector, which is basically a photon detector. It is designed so that any photons detected and transmitted through its system produce and external voltage at some level relative to the number of photons in each stream. These detectors are of limited value in infrared astronomy unless they can be cooled to operate at specific temperatures. Otherwise, the heat from the telescope itself interferes with infrared detection. Liquid nitrogen or liquid helium is generally used to bring photovoltaic detectors to efficient working temperatures.

A second instrument is the photoconductive detector, which is no more than an electronic device that increases its conductivity every time infrared radiation meets the detector's interacting layers. These interacting layers consist of a semi conductor and electrical contacts. (The design is also used for visual and ultraviolet studies). These photoconductive detectors are generally used for detecting radiation at wavelengths running from about 8 to 200 micrometers. These detectors must also be cooled to efficient operating temperatures.

A third of these instruments, used to compensate for the usual photographic-related problems in infrared astronomy, is the balometer. The balometer is used for detection of radiation not only in the infrared region but also in the ultraviolet, just as in the case of the photoconductive detector. A balometer is little more than a resistor designed for absorbing ultraviolet and infrared radiation (and sometimes microwave radiation). It operates on the theory that the more radiation that is present, the higher will be the temperature recorded.

The largest infrared telescope in existence (at this writing) is the 3.8 meter telescope on Mauna

Kea. Mauna Kea is a dormant volcano almost 14,000 feet high, so its 3.8 meter telescope is poised well above much of Earth's atmosphere in a supreme location for infrared studies. The telescope was constructed by the United Kingdom and shares Mauna Kea with other infrared telescopes belonging to the United States (3 meter), and Canada, France, and Hawaii (who jointly own a 3.6 meter). (There are other telescopes on Mauna Kea but these are for radio and optical studies.)

Mauna Kea and other infrared telescopes offer advantages to astronomers. There are a great many objects in the sky that are surrounded by great clouds of gas and dust and, therefore, cannot be detected by optical means. Infrared radiation, however, has the capability of penetrating these great clouds and continue on toward the Earth's atmosphere, where—unfortunately for astronomers but fortunately for the public at large—just about all of the radiation is absorbed. Because they are designed to detect the more heat-intensive radiation, which is exactly what infrared radiation is, infrared telescopes have no daylight limitations. They can be used just as effectively at night as they can be during the day.

The best way to scan the heavens for infrared signals is to get *above* the Earth's atmosphere. And the United Kingdom, the Netherlands, and the U.S.A. have done just that. In the early 1980's, these countries made a joint and successful launching of a satellite specifically designed for infrared studies. Known as the Infrared Astronomical Satellite, by the time it became inactive in 1983—ten months after launching—it had come as close as possible to surveying all of the heavens at wavelengths ranging from 10 to 100 micrometers. The discoveries made as a result of the mission are extensive. The satellite located the possible location of another solar system probably containing worlds that could evolve into planets like Venus or Earth; discovered countless objects never before observed, including comets; and detected many infrared sources that might be newly-emerging stars.

## COSMIC SOURCES

Infrared sources include planets, nebulae, stars, the Galaxy, and other galaxies. As the Sun is a star, and one that gives off a tremendous amount of heat, it is a rich source of infrared emissions. We can feel the infrared rays from the Sun because of its close proximity to Earth. We cannot, however, feel the infrared rays from other stars because they are light years away. In fact, the nearest stars, Alpha Centauri and Proxima Centauri are roughly 1.3 parsecs away. Nebulae and galaxies are also at incredible distances from Earth so their infrared emissions also cannot be felt, and cannot be sensed at all except through the use of highly sophisticated instruments.

Although the planets are relatively close to Earth—we've already sent probes to all but Neptune and Pluto—they are relatively cool worlds and their infrared radiation is not easily picked up. Still, infrared astronomy has helped astronomers discover some things about our solar system. As a matter of fact, the rings of Uranus were first detected by near-infrared observations.

That leaves only the Moon to be added to our list of infrared sources. What we call the Moon is the only one of the Earth's natural satellites still around, and the only one to have been the result of gravitational capture rather than Earth debris. The Moon's infrared signals are fairly easily detected because the Moon is so tremendously close to Earth.

Infrared images are heat images, and heat images are sometimes fairly close outlines of what we would ordinarily see as a result of visual light radiation. So, the infrared sky that the astronomer observes is little altered from the type of view one

**Table 7-1. Infrared Astronomy.**

| System | Major Sources |
|---|---|
| Galactic | Protostars, stars, cool molecular clouds |
| Extragalactic | Seyfert Type II galaxies, BL Lac objects, quasars |

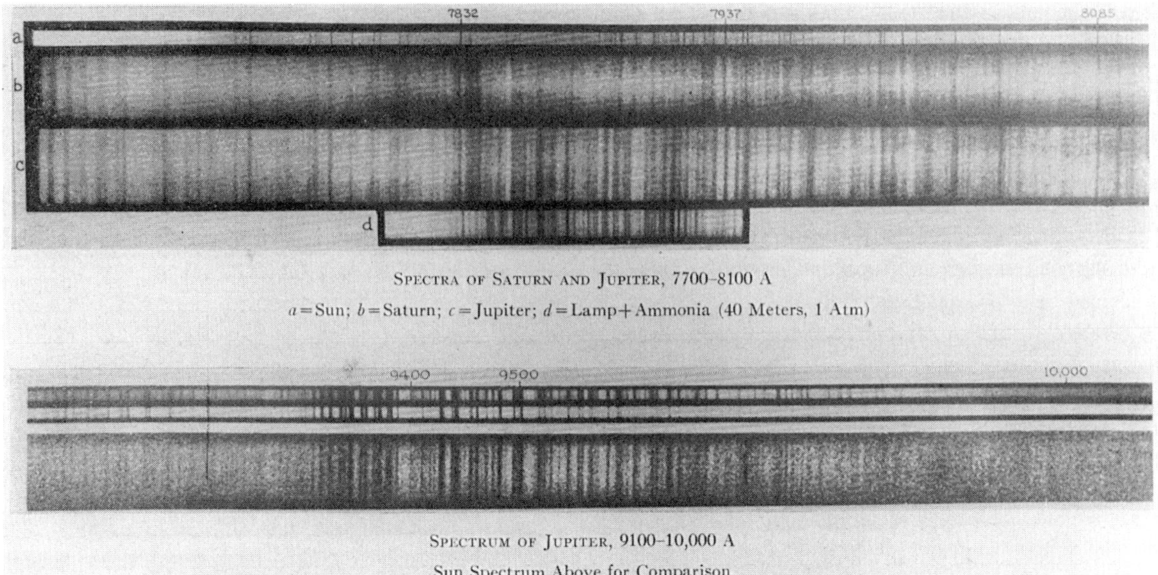


Fig. 7-1. The infrared spectra of Saturn and Jupiter indicates the presence of ammonia in their atmosphere (Courtesy of Yerkes Observatory, from Kuiper, *Atmospheres of the Earth and Planets*, 1949).

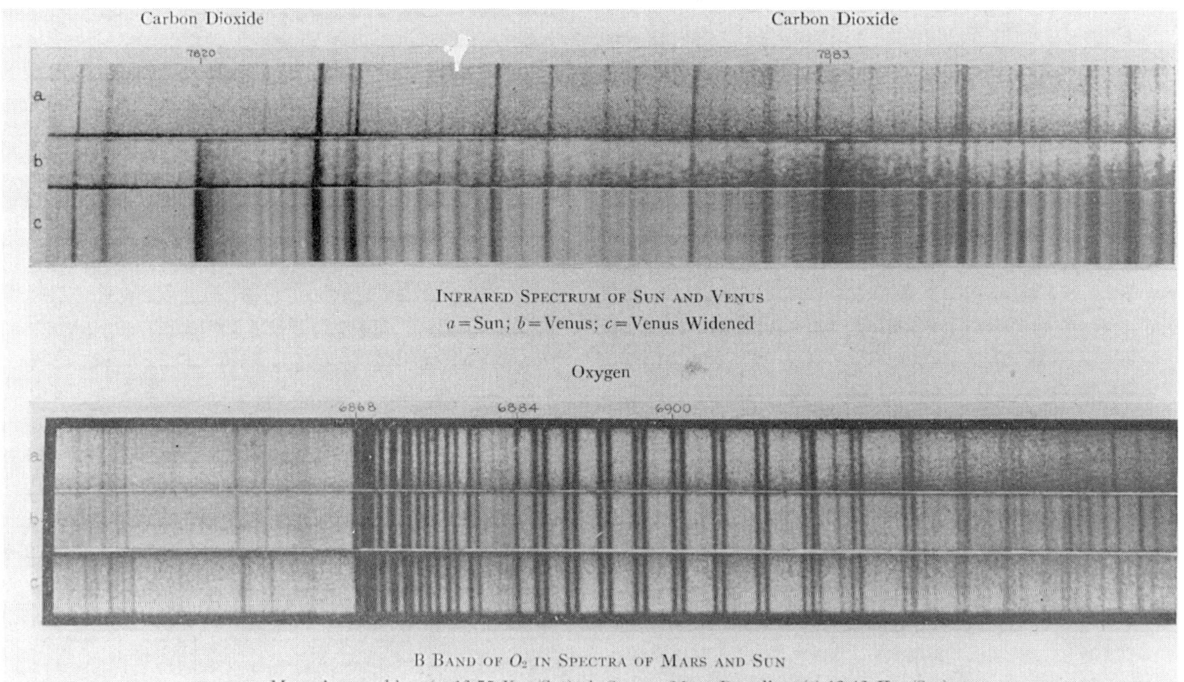


Fig. 7-2. A comparison of the infrared spectra of Venus and Mars with the spectrum of the Sun, illustrating the existence of carbon dioxide in the atmosphere of Venus and the lack of oxygen in the atmosphere of Mars (Courtesy of Yerkes Observatory, from Kuiper, *Atmospheres of the Earth and Planets*, 1949).

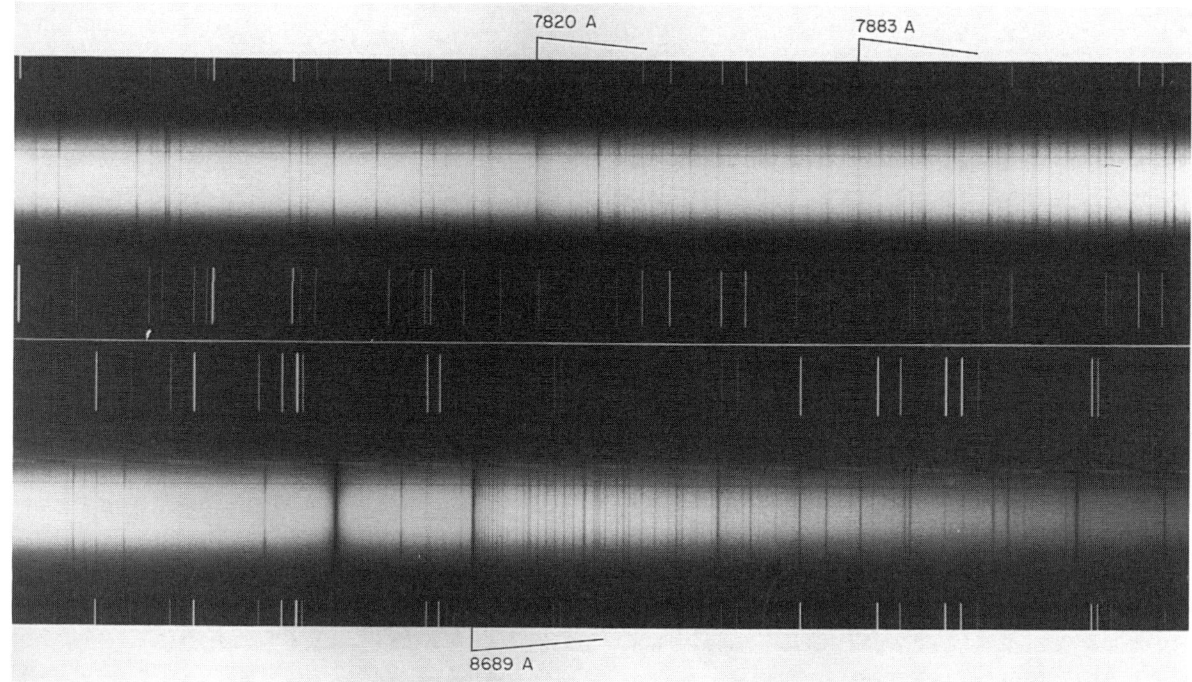


Fig. 7-3. Spectrum of Venus, showing infrared carbon dioxide band (Courtesy of Lick Observatory).

would expect from visual light, although color and brightness are necessarily altered and the hottest blue stars (Type O) come across as rather dim sources in infrared and the cooler red stars (Type M) come across as rather rich sources. (See Table 7-2). To get an idea of the way in which infrared images are detected, place your warm hand against a cold window, and you will see a heat impression of your hand left on the glass; the hotter your body temperature, the more profound the image.

The greatly distant and highly luminous galaxies that are currently emitting strong infrared waves are suspected of actually being quasars. Now, quasars generally emit at optical and ultraviolet wavelengths, so why do astronomers suspect a connection? Well, these galaxies seem to be producing about the same amount of energy that astronomers usually detect coming from quasars, except that in the case of these highly luminous galaxies almost 100 percent of the emissions appear in the infrared regions. This infrared radiation is probably the result of the galaxies' absorption of optical radiation and their re-emission of these optical waves as infrared waves.

**Table 7-2. Spectral Type and Temperature Ranges of Stellar Objects.**

| Type | Color | Temperature Range (Kelvin) |
|---|---|---|
| 0 | Blue | 28,000-40,000 |
| B | Blue | 10,000-28,000 |
| A | Blue & Blue-White | 7,500-10,000 |
| F | White | 6,000-7,500 |
| G | Yellow | 5,000-6,000 |
| K | Orange-Red | 3,500-5,000 |
| M | Red | -3,500 |

## PROTOSTARS AND RED GIANTS

Some stars do not emit infrared rays in the relation one might expect to their spectral class. That is, many of the cooler stars show strong emissions at certain wavelengths other than normally expected. This phenomenon results from the heating of the dust clouds by ultraviolet emissions coming from the star being investigated in the infrared.[1] The source of the emissions, the star within the dust clouds, is probably a very young star, probably one still in the protostar stage, for stars like our Sun—middle-agers in the cosmic time scale—have no dust clouds protecting or hiding them.

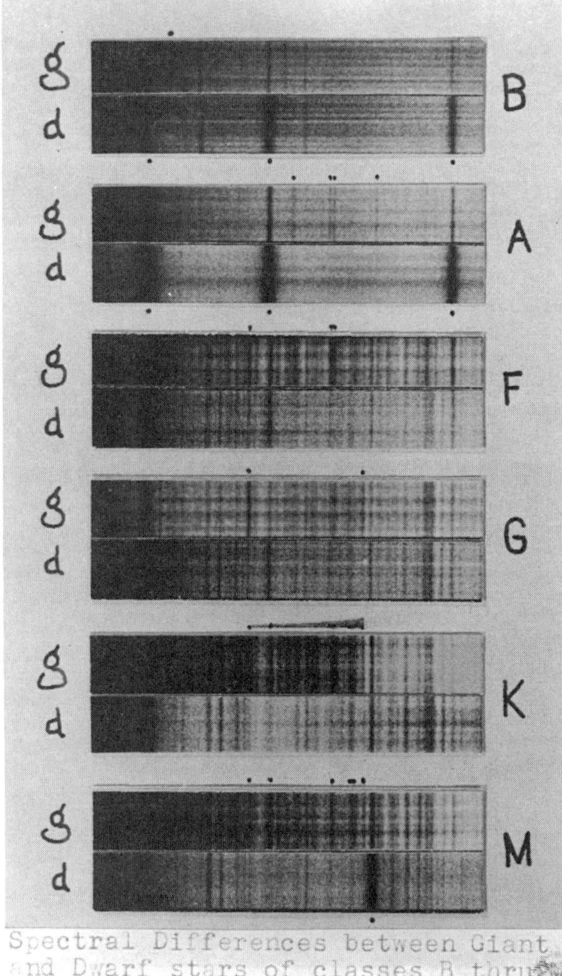


Fig. 7-4. The spectral differences between giant (top) and dwarf stars (Courtesy of Yerkes Observatory).

To understand this, it is necessary to have a little deeper understanding of the way in which stars come into existence—or, rather, to understand the way in which astrophysicists *theorize* about the way in which stars are born. No one can be certain of the stellar birth and evolutionary cycle, because no one has been around long enough to chronicle the birth, childhood, adolescence, middle age, old age, and death of a star.

As the theory goes, stars are born in molecular clouds (described in the last chapter)—not every star, of course, but probably the vast majority of them. As these clouds collapse, possibly as the result of a supernova explosion parsecs away, instabilities caused by the effect of the explosion cause the original dust formations in the cloud to become fragmentized. Each of these fragments becomes a subcloud, which, in turn, continue to divide just as did the mother cloud. This fragmentation process cannot go on indefinitely. At some time the process must end and it does so when some of the sibling clouds acquire so much density and heat that physical laws prevent any further divisions.

These highly dense fragmented clouds now enter into the protostar stage. This is the stage of a star before it has collapsed enough to allow the kind of nuclear reactions that will make it a potent energy field. For a star the mass of our Sun, the cooler part of the protostar stage—the time before the nuclear activity—may last up to a million years or more. After this time, the "baby" star begins to get ever more dense and ever more hot until the nucleus of the star is formed. Meanwhile, the star is still surrounded by remnants of the cloud which gave it birth. Now, as the star grows even hotter, it burns away the surrounding dust and starts life on its own. Until it is burned away, this dust cloud, heated by the ultraviolet rays from the protostar, continues to give off the infrared emissions that astronomers detect. What appear to be infrared stars are not stars at all but rather cosmic dust clouds surrounding newly emerging stars.

Table 7-3. Stellar Evolutionary Theory.

| Stage | Description |
|---|---|
| I: Cloud Collapse | A cool dense cloud of gas and dust collapses, and in so doing is separated into fragments. |
| II: Cloud Fragmentation | Fragmented clouds become too hot and dense for further division and so begin now to form protostars. |
| III: Protostar Formation | Protostar collapses until a very hot and very dense core develops which serves as the nucleus for the newly emerging star. |
| IV: Star Formation | The core heats to a temperature of about 10 million degrees Kelvin and now the nuclear reactions begin which, in effect, complete the object's transition from protostar to main sequence star. |
| V: Main Sequence | Star continues to produce energy through hydrogen to helium conversion during a period in its evolution which may last billions of years. |
| VII: Hydrogen Exhaustion | Main hydrogen supplies have become exhausted and the star's core begins to collapse. A hydrogen shell about the star now serves as fuel. |
| VIII: Red Giant | Temperatures have settled around 3000 degrees Kelvin and the diameter of the star is probably 100 times that of the Sun. This red giant will go through periods of contraction and expansion as its interior physics is seriously altered due to nuclear activity. |
| IX: White Dwarf | Star finally collapses into a white dwarf (of less than 1.5 solar masses) and continues to burn away over a period of, perhaps, ½ billion years. |

The clouds in the infrared sky are not necessarily those that surround new stars. They may also represent the environment of stars which are nearing the end of their life spans (of millions or billions of years). When a star with a mass similar to that of our own Sun is dying, it has depleted its main supply of hydrogen and is now beginning to collapse. The collapsing continues, and as it does temperatures of the star increase, until the environment becomes so intensely hot that a shell of burning hydrogen comes to envelope the core of the star. The star is now forced to expand and as it does so, it begins to cool and becomes what is called a *red giant*, a star that can be more than 100 times the size of the Sun but only 2500 degrees Kelvin in temperature. The star may go through another period of contraction for a period of hundreds of thousands or millions of years, only to expand once more to the red giant stage. In any event, during its expansion stage (or stages) the dying star "bleeds" gas into space and atomic structures within the gas tend to clump into minute dust grains which form a great cloud about the star. The heat from the star, as you have now guessed, heats this new cloud and generates the infrared emissions detected by astronomers.

Not all stars end in this way. Those which have more than twice the mass of the Sun take a different and quicker evolutionary route. The stars generally exhaust their core hydrogen and then begin to burn away their helium until they then begin to contract to such an extent that their carbon content is converted into oxygen, neon, and magnesium, now setting up the conditions that excite a supernova explosion. And, as explained earlier, the results of a supernova explosion turn a star into either a neutron star or a black hole.

## OH/IR STARS

Still another source of infrared emissions is a population of celestial worlds called OH/IR stars.

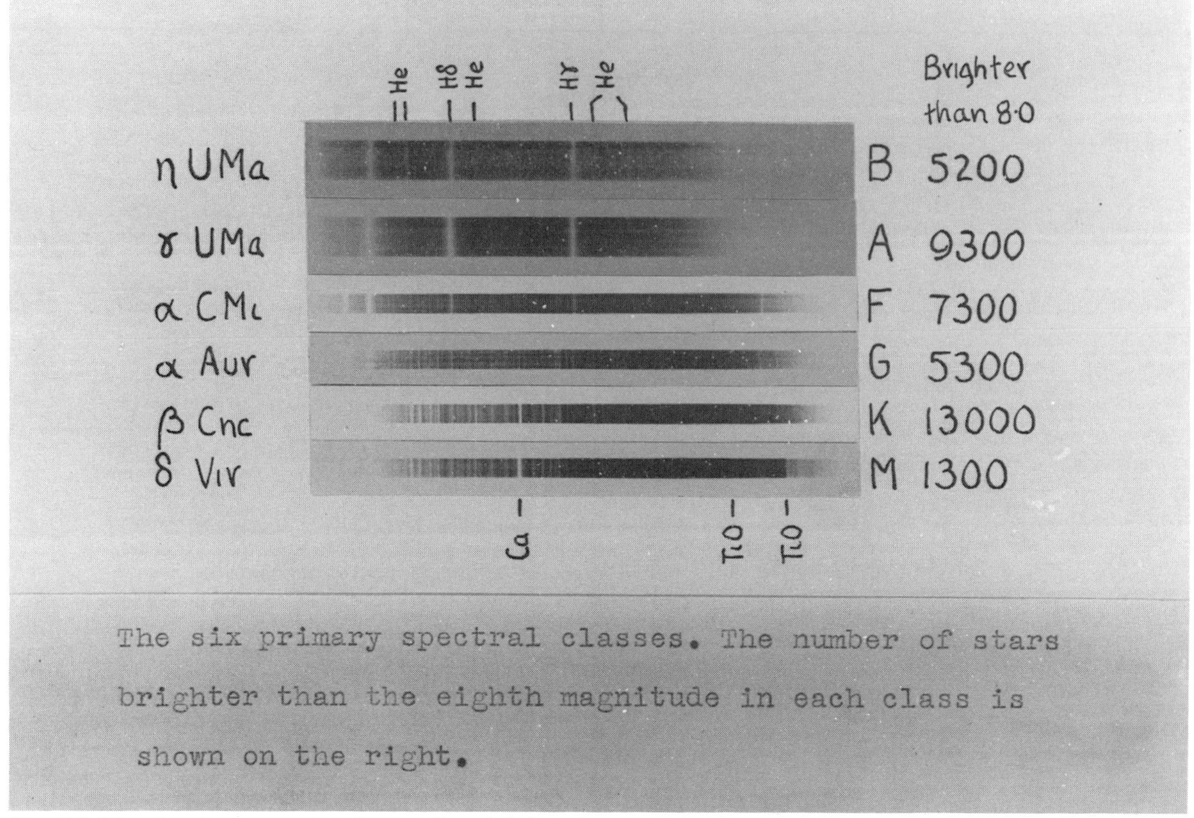

Fig. 7-5. The six primary spectral classes. Stars brighter than the 8th magnitude in each class are shown on the right.

These are among the largest and brightest stars in the sky, so therefore qualify as either giants or supergiants. The supergiants are the more rare members of the stellar population so it stands to reason that most of the OH/IR stars are giants, those highly luminous stars in the latter phase of their evolution that have extremely dense centers but not-so-dense atmospheres.

OH stands for the hydroxyl radical[2] and IR stands for infrared. OH/IR stars are identified by their hydroxyl maser emissions and their infrared radiation. Masers are celestial objects serving as a source of electromagnetic radiation caused by excited molecules or atoms. There are actually a number of types of masers but the OH group is only of concern here.

OH/IR stars were detected in the early 1970's but the vast majority of those catalogued today were not discovered until a decade later, and only because of the combined efforts of the Netherlands, the United Kingdom and the United States through their infrared satellite program (IRAS). OH/IR stars are very quickly deteriorating and are engulfed in clouds of dust with temperatures ranging from tens of hundreds of degrees kelvin. Before long, the OH/IR worlds will be little more than the dust clouds which now surround them.

There is another type of star, almost classified as an OH/IR, that also gives off strong infrared emissions. This type of star is named after a variable star known by, and popular with, the ancient astronomers for its exceptional brightness. The star is Mira, a red giant in the constellation Cetus; long period variables named for it are known as the Mira Ceti variables, or Miras for short. Miras are relatively numerous but they differ considerably from the hydroxyl maser infrareds in terms of the strength of their radiation. The OH/IR stars are much, much stronger maser and infrared sources than the Miras.

## THE GALACTIC CENTER

The Sun leads Earth and the other planets in our solar system on a perpetual circular tour around

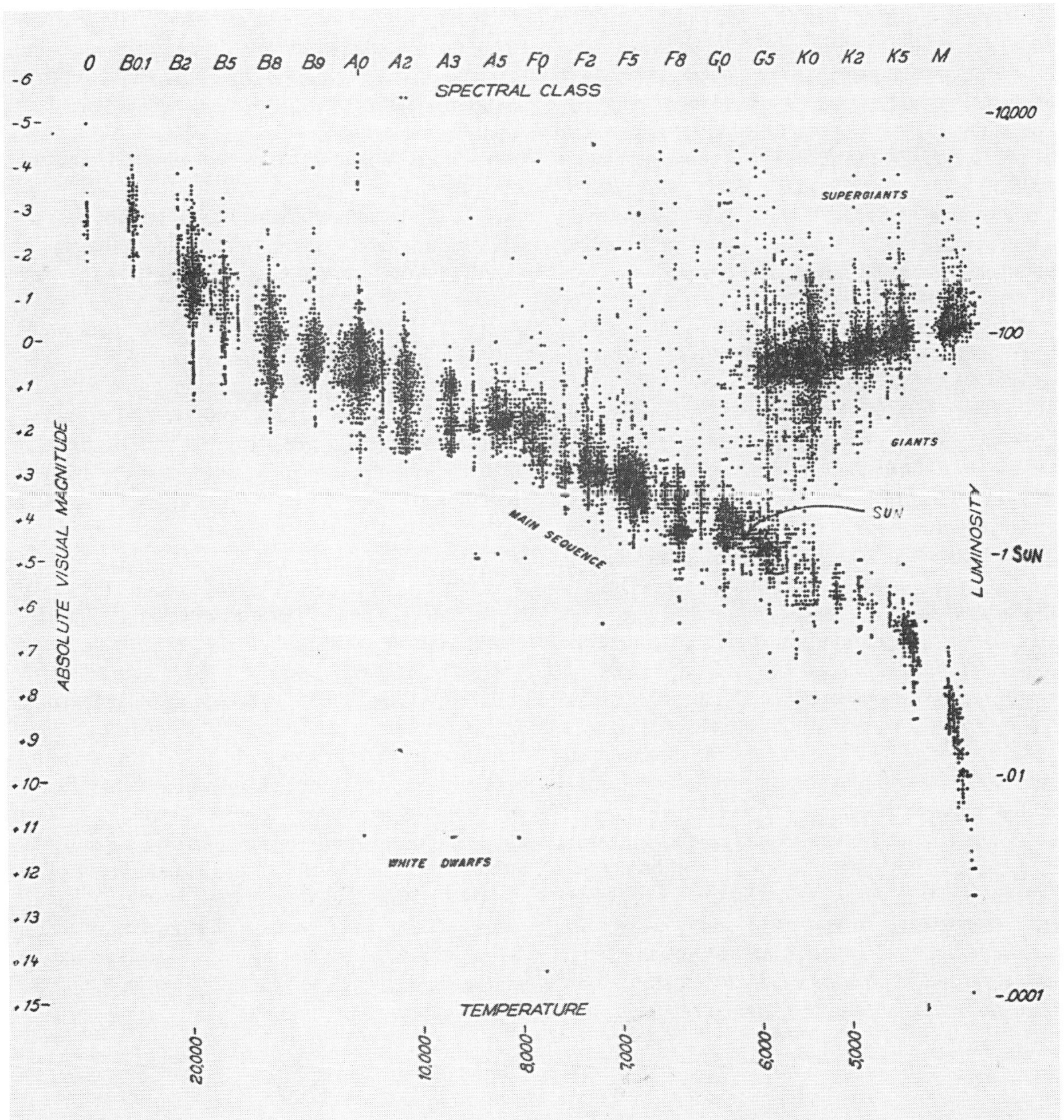


Fig. 7-6. The Hertzsprung-Russell diagram, plotted to show the absolute magnitudes of stars against their spectral types (From Struve, *Stellar Evolution*, Princeton U. Press, 1950, courtesy of Yerkes Observatory).

the center of the galaxy. The galactic center is roughly 10,000 parsecs away in the direction of the constellation Sagitarius. Sagitarius is a southern constellation symbolized by a centaur aiming a bow and arrow.

Actually, it is not just the Sun which is circling about the galactic center. Every object in the galaxy is on a similar journey. From a spot somewhere in eternity, or in the great cosmic beyond—whatever you many want to call what lies beyond the realm of our universe—the galaxy looks like a spinning disk. The core about which the galaxy's components circle is covered by blinding clouds of gas and dust. There is no possible way that visual radiation can penetrate this great cover. But infrared rays can, as well as radio, X-ray, and gamma rays. Astronomers in each of these fields have joined forces to try and determine what lies at the center of the galaxy. Radio findings were discussed in the previous chapter, and X-ray and gamma ray research will be discussed in coming chapters.

Infrared detection has turned up the existence of a yet unidentified object or system lying about 35 or 40 parsecs above and below the galactic plane. The galactic plane is the hypothetical plane containing the galactic equator that almost passes through the central plane of the Galaxy's spiral disk.

This strange entity, extending for parsecs and beginning not far from the galactic plane, is shadowed by clumps of celestial matter which astronomers believe indicate the presence of O-type and B-type stars. O-type stars are very hot and massive blue stars with temperatures possibly reaching 50,000 degrees Kelvin.

Generally, O-stars are found in the spiral arms of galaxies. They rotate relatively swiftly and live for only a million years or so while most stars continue in existence for billions of years. B-stars are also very fast rotators; they are distinguished from hot, main-sequence stars called "shell stars" only by minor differences in appearance resulting from the variations to be expected when objects are examined from different angles.

The only real difference between shell stars and B-stars is related to the angles of observation. If these stars are generally found in the outer parts of the galaxy or in spiral arms, what makes astronomers think that these strange clumps near the galactic core are O's and B's? These clumps appear to be producing infrared and radio signals with relatively equal intensity. This is a characteristic of O-type and B-type stars, thus the reason behind the speculations.

These O and B stars might also be the reasonable explanation for that extended object the size of hundreds of thousands of O and B stars. Astronomers consider that if there were indeed a great conglomeration of these stars (O and B), they would probably be shielded by a great cloud of fairly even density throughout. The cloud would give off emissions in the infrared ranges, but give off the energy in such a way as to leave observers with the impression that one great object is producing the infrared radiation.

## COMETS

Comets give off infrared emissions. The Infrared Astronomical Satellite has recently detected narrow trails of dust that are probably caused by comets. In a May 1986 issue of *Science*, a joint article by researchers at the University of Arizona explains that as one of the known periodic comets makes its way closer to the Sun, the solar wind dislodges debris from the nucleus of the comet and blows it along the wake of the comet. The debris then follows an orbit very similar to that of the comet. This trail of debris is compounded as cometary passages continue, so that over periods of hundreds of years there is quite a buildup of loose material, and an energy field strong enough to produce detectable infrared emissions is created.

# Chapter 8
# Ultraviolet Astronomy

A YOUNG CONTEMPORARY OF HERSCHEL'S, JOHANN Wilhelm Ritter (1776-1801), in his very short life, illustrated his genius in just about everything he ever undertook. Once a pharmacist and also a teacher of medicine, Ritter soon realized that his first love was physics, and he was especially interested in the subjects of electricity and electrochemistry. His contributions to science were impressive by any measure. He discovered the process of electroplating, invented the dry voltaic cell and the electrical storage battery, and did important research with thermoelectric currents.

The year after Herschel discovered and named infrared radiation, Ritter devised an experiment similar to Herschel's. But Ritter directed a beam of solar light into dampened silver chloride instead of just separating the beam for temperature analysis. As expected, he discovered that the color of the light-sensitive compound changed according to which part of the spectrum was affecting it. However, he found that the greatest effect was not caused in the observable ranges of the spectrum but by some invisible field on the other side of the violet ranges.[1] And so came the discovery of ultraviolet radiation.

Though it was almost two centuries ago that Ritter discovered the existence of ultraviolet radiation, it was not until very recently that ultraviolet astronomy came into being. Actually, ultraviolet astronomy is a product of the space age.

Ultraviolet radiation occurs in the 4-400 nanometer wavelength range. It is generally divided into three limits, or wavelength fields, which are called the near, far, and extreme ultraviolet regions.[2] Near ultraviolet takes up the 400-300 nanometer range, for ultraviolet the 300-200 nanometer range, and the extreme ultraviolet the 200-4 nanometer range. Frequency ranges for ultraviolet radiation is from roughly $10^{15}$ to $10^{18}$ hertz. For the most part, ultraviolet rays from celestial objects cannot make it to the surface of Earth because of the atmospheric presence of ozone, oxygen, and nitrogen. Thus, even more than the infrared astronomer, the ultraviolet astronomer must take his instruments as high above the surface of Earth as possible. Satel-

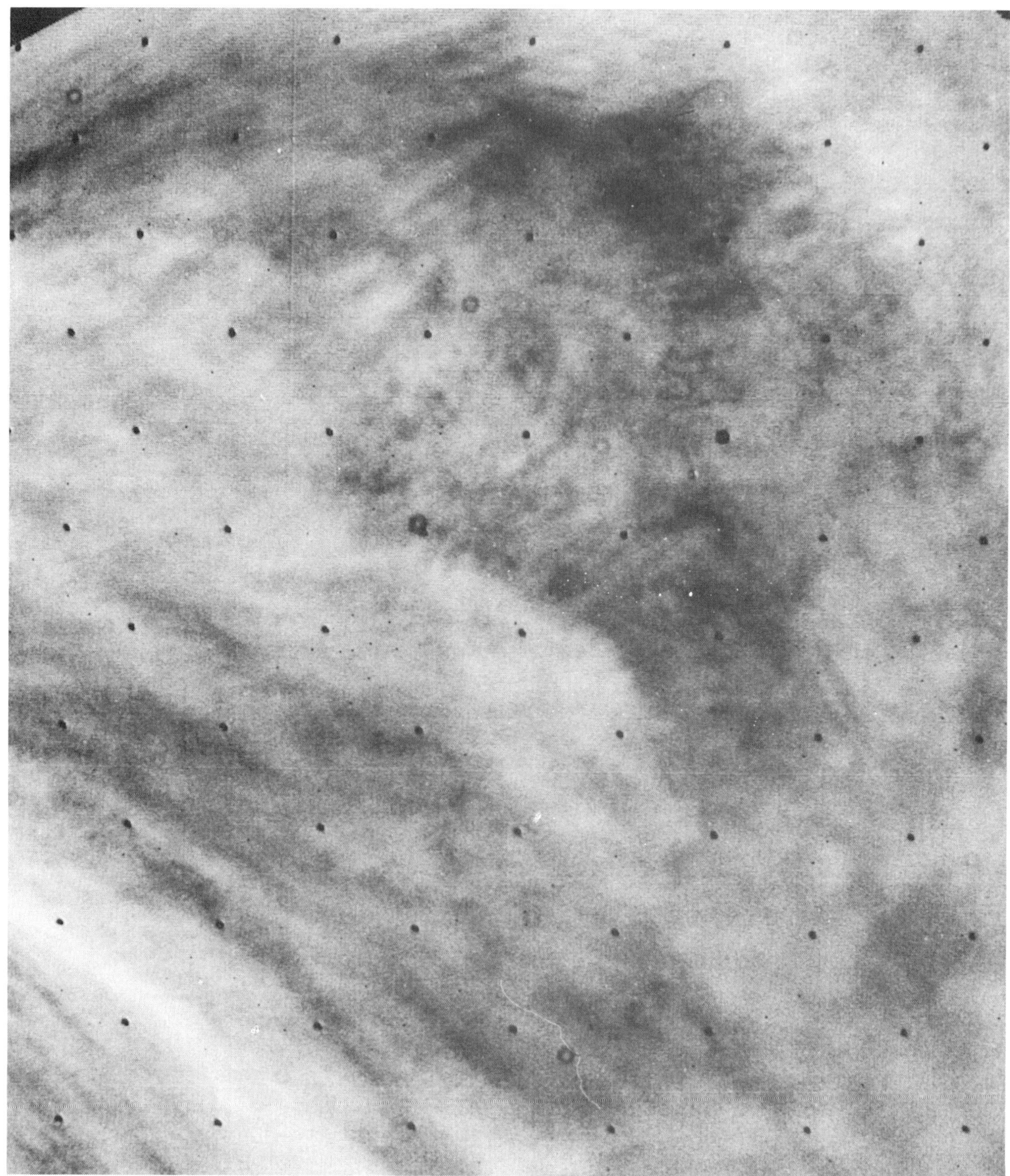

Fig. 8-1. This ultraviolet photo of Venus was taken in 1974 by Mariner 10's television cameras from a distance of about 490,000 miles. Dark features near the top are a section of a dark cloud belt over the equitorial region. There are rising and descending air currents quite typical of convection on Earth. To the south of this cloud belt are spiral-like streaks which indicate a uniform flow around the planet toward the pole (Courtesy of NASA).

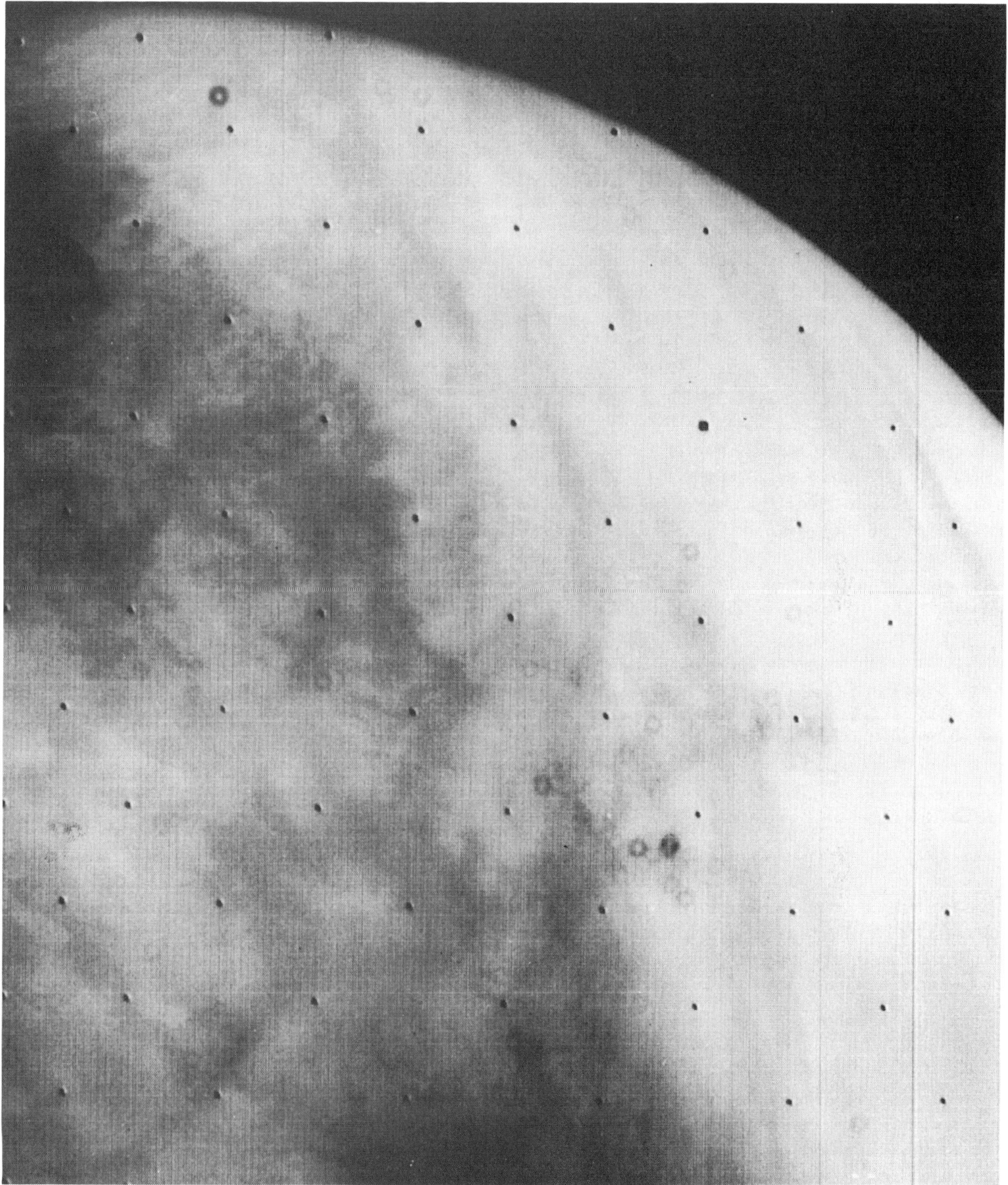

Fig. 8-2. This ultraviolet photo of Venus was also taken through ultraviolet filters. The dark belt at the bottom straddles the equator and, again, detail within the belt shows rising and descending air currents quite typical of convection on Earth (Courtesy of NASA).

lites, space probes, and rockets usually serve as ultraviolet astronomy observatories. Near-ultraviolet rays can be detected by ground-based observatories using optical and radio telescopes.

Rocket astronomy has played a particularly important role in solar studies, particularly for the detection of radiation that is generally absorbed at altitudes between 240 and 40 kilometers. Rocket astronomy got its start in the mid-forties as the result of the capture of German U-2 rockets. In Germany, about twenty years before World War II, a group of enthusiastic amateur astronomers got together to found a group dedicated to the research and development of rockets or other vehicles that could be launched into space. The members called themselves the Society for Space Travel (Verein für Raumshiffahrt). However, when they attempted to register themselves as a professional organization, they were openly ridiculed. Government authorities explained that they were dreamers, men who were chasing after the impossible. Authorities cited the intended name for the group, "Raumshiffahrt;" it meant travel by spaceship. And the idea of travel by spaceship, the authorities further explained, was not even a fathomable idea. Besides, Raumshiffahrt was a contraction of two separate German words (spaceship and travel) that were not used together in any dictionary and which were not even in use together by the scientific community.[3]

## SPACE BASED OBSERVATORIES

Because near-ultraviolet radiation is able to make it through Earth's atmosphere and down to the surface, ultraviolet observations at the 400-300 nanometer range require no special technology other than that used for standard optical observatories. The same telescopes and the same photographic techniques can be employed. Special problems are presented when far and extreme ultraviolet observations are to be conducted. Some of these problems are related to the special characteristics of ultraviolet rays, so the telescopes used must be built with mirrors designed with special coating to ensure the degree of reflection required. Most of the problems have to do with the need for high-altitude observations.

No mountain on Earth is high enough for far and extreme ultraviolet observatories, but astronomers need not always employ rockets and satellites for their research. In some cases, balloons can be used to altitudes of around 30 kilometers where ultraviolet wavelengths in the 250-300 nanometer range can be detected. As astronomers always want to maximize their efforts and have as much of the ultraviolet spectrum to survey at any one time as possible, they carry out most of their research from man-made satellites.

The first use of satellites for ultraviolet astronomy dates back to the late 1960s and to the launching of a series of satellites collectively called the Orbiting Astronomical Observatories (OAO's). The launchings continued through 1972 and were highly successful experiments in the cataloging and photographing of ultraviolet sources and in the measuring of ultraviolet luminosity in selected fields. Stella spectral data in wavelength ranges from 300-90 nanometers was gathered, resulting in the compilation of a great deal of data about interstellar clouds and stellar winds.

In 1972, the European Space Research Organization launched the TD-1 satellite, also designed for ultraviolet observations. Its mission, quite efficiently accomplished, was to make an ultraviolet survey of the entire sky. Two years later, the Astronomical Netherlands Satellite was launched, carrying not only an ultraviolet telescope designed to provide broad-based photometric data at selected wavelengths, but also two X-ray telescopes. In the two years that the satellite was in operation, it made tens of thousands of ultraviolet observations in the near-ultraviolet to extreme ultraviolet regions.

In 1978, the European Space Agency, the United Kingdom's Scientific and Engineering Council, and NASA launched their International Ultraviolet Explorer, a satellite carrying a 45-cm telescope specially designed for ultraviolet detection in the 115 to 320 nanometer lengths. Perhaps, by the time this book is published, the satellite will have run its course and become inoperable, but if its past success is any indication of future success, this international venture will have succeeded in obtaining spectral data on 80,000 to 100,000 celestial objects, from comets and planets to stars and galaxies.

In any event, by the time the International Ultraviolet Explorer is no longer operable, NASA hopes to have in orbit its new space telescope (Hubble Space Telescope). This is a 2.4 meter telescope almost 95 feet in length. It is capable of ultraviolet, optical, and near-infrared studies for a period of about 15 years. By the turn of the century, thanks to the assistance of the European Space Agency, NASA will have come as close as possible to mapping everything detectable in the sky—and will probably bring about more constructive cosmological models of the universe.

## SOLAR SOURCES

Major ultraviolet observation of the Sun began with the Skylab missions in 1973. These missions were designed to detect both X-ray and ultraviolet spectral regions. High above Earth's atmosphere, Skylab could study the Sun's corona and chromosphere with fewer restrictions than astronomers were used to.

Skylab actually consisted of one great space station designed as a floating city in the sky that astronomers could visit to conduct research and make whatever repairs were necessary to the space station. Skylab 1 was the space station.[4] Skylab 2-4 were the manned missions to the space station. The space station contained an Apollo telescope mount, or ATM, which housed the necessary instruments for X-ray and ultraviolet observations. Among its instruments were two X-ray telescopes and a white light coronograph;[5] an XUV spectroheliograph, an ultraviolet spectroheliometer and spectrograph; and two hydrogen alpha telescopes. The X-ray equipment was used to study the solar corona; the ultraviolet equipment was used to study the chromosphere; and the hydrogen-alpha telescopes, which operated at 656.3 nanometers, were used to observe the lower chromosphere.[6]

The chromosphere is the lower part of the sun's atmosphere. It is a wall of hydrogen gas that is perhaps six thousand miles or so thick. From Earth, it is absolutely impossible to see the chromosphere without special astronomical equipment, unless the photosphere—the highly visible and extremely luminous solar surface—is totally eclipsed by the Moon. At this time, the chromosphere appears as a very thin and reddish field forming what might look like a quarter-moon. Its light always seems to be flashing as the Moon continues to cross past the Sun, the very brightest flashes being the result of an overabundance of calcium, helium, and hydrogen.

The corona is a gaseous envelope of rarefied gas that makes up the outer part of the Sun's atmosphere. It is divided into three fields labeled K, F, and E. The K-corona and F-corona comprise the white-light portion. The K-corona is the most intense field and reaches a temperature of about two million Kelvin. The F-corona is the outer part of the corona. The E-corona is that part of the corona which is made up of slow moving ions.

As is the case with the chromosphere, the white light fields of the corona cannot be seen except during an eclipse. The very first white light coronal images, therefore, had to be made from space, and rocket astronomers were the first contributors. Analyses of these first white light images, obtained in the 1960s and early 1970s, showed that the outer corona goes through a series of extreme changes over periods as short as a day. Observations via rocket also provided data verifying that activity in the corona can result in magnetic storms on Earth, this occurring when plasmatic masses are hurled into space at speeds in excess of 1000 kilometers per second.

Since the Skylab missions, a great deal of additional information has been compiled about this great star that gives both life and light to us. The Sun remains a great subject for ultraviolet studies because it is so close and its rays so intense. By analyzing the ultraviolet emissions from this closest star, astronomers are able to determine the physical and chemical properties of the Sun and help to fill in pieces to the puzzle of its birth and evolution.

Ultraviolet studies create a picture of the Sun that give it the appearance of a great ball of fire spit up from the bowels of a volcano. This is exactly what visual light studies paint the Sun to be. And why not? After all, this is exactly what the Sun is, a great world of fire, a world of unimaginable temperatures and constant and great explosions.

Fig. 8-3. The two-stage Saturn rocket carrying the Skylab 4 astronauts lifts off from the launch complex on November 16, 1973. This was the third of Skylab's manned missions (Courtesy of NASA).

Fig. 8-4. An overhead view of the Skylab space station. The station contained an Apollo telescope mount, X-ray telescopes, an XUV spectroheliograph, hydrogen alpha telescopes and other detecting equipment. X-ray equipment was for study of the solar corona and the ultraviolet equipment was for study of the upper chromosphere (Courtesy of NASA).

Fig. 8-5. Skylab 4 command module carrying the Skylab astronauts Gerald P. Carr, Dr. Edward G. Gibson, and William R. Pogue splashes down in the Pacific February 8, 1974. The Skylab missions were designed to collect X-ray and ultraviolet solar data (Courtesy of NASA).

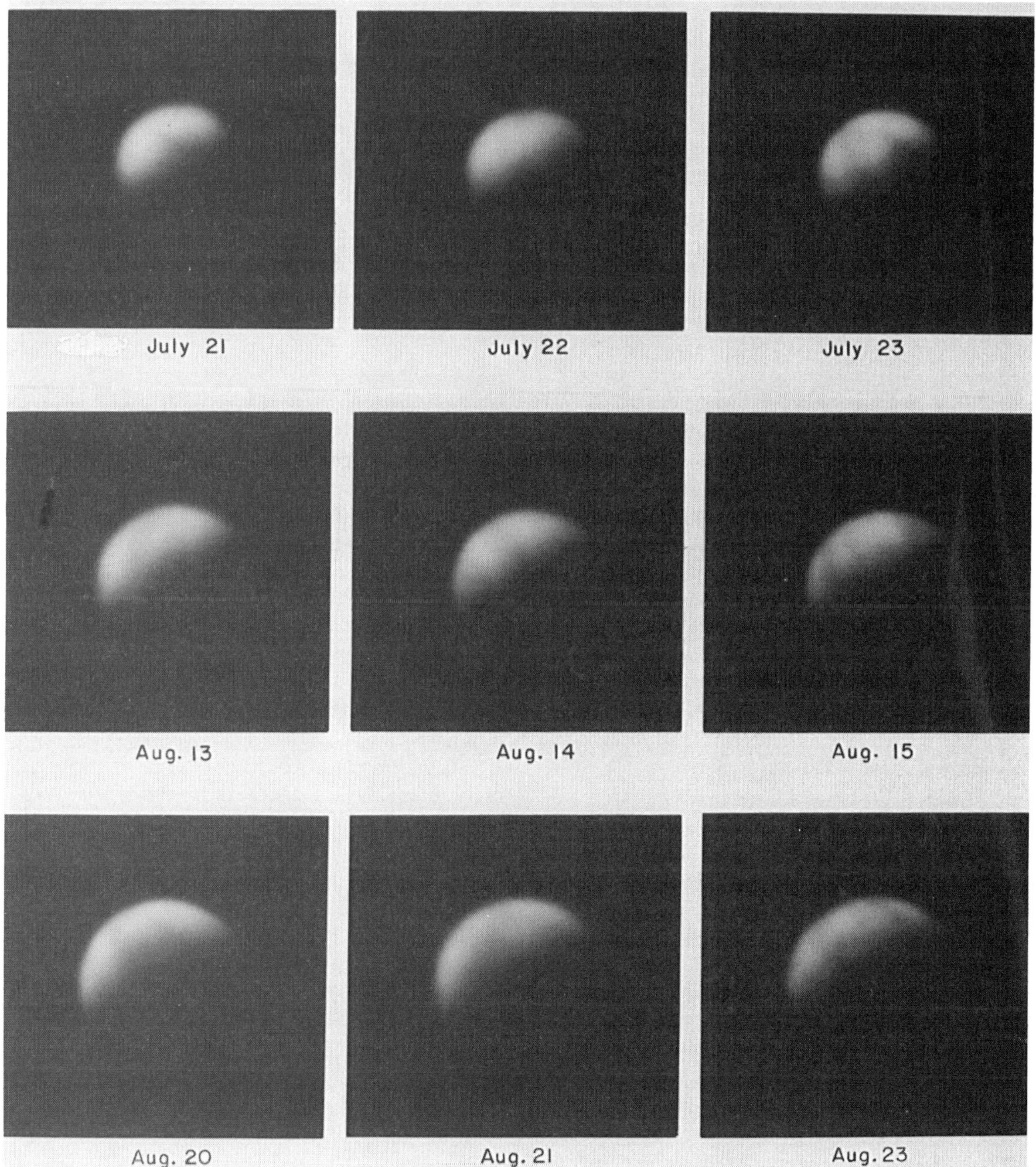


Fig. 8-6. This combination photograph of Venus taken with the Lick Observatory 120-inch reflector shows Venus and its cloud markings in ultraviolet light.

Specifically, the Sun is a world made up predominately of hydrogen and helium in about a 3 to 1 ratio. Most of its energy is produced by nuclear fusion, more specifically by a proton-proton chain reaction. In this type of nuclear fusion, requiring a temperature of $10^7$ degrees Kelvin, energy is produced through the conversion of hydrogen to helium. The temperature near the core may be as much as 20 million degrees Kelvin but it cools rather rapidly until it falls to about 6,000 degrees at the surface. At this temperature (6,000 degrees), the Sun qualifies as a G (yellow) star and as such, if it were further away, it would hardly be much of an ultraviolet source.

The extent to which ultraviolet studies of the Sun will contribute knowledge about other stars is still to be determined, for ultraviolet astronomy is such a new science. There are many stars which radiate to Earth mainly in the ultraviolet regions of the spectrum, and so all that can be gathered about them depends mainly on this ultraviolet profile. By comparing their ultraviolet images with that of the Sun, a lot can be determined about these stars. This means of stellar research is bi-directional. As astronomers learn more about other stars through the other fields of astronomy as well as through ultraviolet studies, spectral comparisons add so much more to what we know about the Sun.

## STELLAR SOURCES

Stars which are under 10,000 degrees Kelvin at their surface generally radiate mostly in the visual wavelengths. This means that A-stars (7,500-10,000 K), F-stars (6,000-7,500 K), G-stars (5,000-6,000 K), K-stars (3,500-5,000 K), and M-stars (less than 3,500 K) are not subjects for ultraviolet study. This means that red, orange-red, white, blue-white and some blue stars, as well as yellow stars like our Sun, will emit little ultraviolet radiation toward Earth. Of course, the proximity of our Sun enhances ultraviolet detection in this case, and space probes that sometime in the future will go beyond our solar system, will probably be able to pick up some measure of stellar ultraviolet emissions.

The stars that make good ultraviolet subjects are the extremely hot stars, like the blue O and B stars. The O-stars have a temperature range of 28,000 to 100,000 degrees Kelvin; the B-stars have a temperature range of 10,000 to 28,000 degrees Kelvin.

The O and B (and possibly A) stars are very hot and massive stars that probably generate up to 80 percent of their radiation at ultraviolet wavelengths. These very hot stars have relatively short lives because more massive stars (than the Sun) convert hydrogen to helium through the carbon cycle rather than the proton-proton cycle. The carbon cycle is a type of nuclear fusion that makes the cores of the stars less dense than the Sun. One of the results of the process is that the star uses up its helium at a tremendously rapid pace and, therefore, grows old much faster than solar mass stars, finally becoming a neutron star or black hole in a period of just a few million years.

Thus, the stars which are roughly equivalent in mass to our Sun, and so have been privileged with life spans that run into billions of years, are not strong emitters of ultraviolet radiation. They represent the more mature members of the universe, have been around for billions of years, and will be around for many billions more. Someday they must burn up most of their hydrogen and begin a phase of contraction that will also greatly increase their temperatures. Eventually that series of nuclear reactions will begin which form a hydrogen shell about

**Table 8-1. Ultraviolet Astronomy.**

| System | Major Sources |
|---|---|
| Galaxy | O stars, B stars, white dwarfs, interstellar space, Magellanic clouds |
| Extragalactic | Seyfert galaxies, quasars |

Fig. 8-7. Coma group spectra of 14 stars, photographed with ultraviolet doublet and objective prism (Courtesy of Yerkes Observatory).

the core of dying solar-mass stars. The Sun will expand to 100 times its present size and become a red giant. When all its fuel is gone and the nuclear activity ends, the Sun will collapse into a white dwarf, roughly 1/100th the size of the Sun today.

In addition to the compilation of data on solar mass stars, like O and B stars, ultraviolet researchers have been able to provide data on much cooler members of the stellar population, by concentrating their studies on specific areas of stellar atmospheres. Such studies have resulted in detailed maps of the subject stars, helping to provide a picture of the way in which temperatures are distributed throughout stellar atmospheres and helping to provide the kind of data astrophysicists need in order to determine the way energy is distributed in different stellar systems.

## WHITE DWARFS

White dwarfs are burned out stars, but this does not mean they are cool bodies. White(normal) dwarfs are very hot stellar remnants. When a star of stellar mass eventually evolves into a white dwarf, it still continues to give off an intense amount of heat, and still emits strongly in the ultraviolet regions. This holds true for subdwarfs, which are stars—or the remnants of stars—even smaller than the main-sequence dwarfs. These subdwarfs are extremely ancient members of the cosmos, have an extremely

Table 8-2. Dwarfs and Giants (The Luminosity Classes).

| Type of Star | Description | Class |
|---|---|---|
| Supergiant | Largest type of star, extremely massive and extremely bright. | Bright Supergiant: I-a<br>Supergiant: I-b |
| Giant | Giants of relatively low mass. In advanced stages of their evolution. | Bright Giant: II<br>Giant:III |
| Subgiant | A star of the giant class but a bit smaller and not quite so bright as other giants of the same spectral class (Table 7-2). | IV |
| Dwarf | Stars about the size and mass of our Sun, though these stars are now more frequently referred to as main-sequence stars. | V |
| Subdwarf | Stars somewhat smaller than the Sun and not quite as bright. They have a low metal content and this effects their temperature and brightness. | VI |
| White Dwarf | A star in the last stages of its evolution. It has already undergone extreme gravitational collapse. | VIII |

low metal content, and are generally much dimmer than normal dwarfs. Being able to distinguish between normal and subdwarfs allows astronomers to compare their number in selected fields. By looking at their ratio to other bodies in the sky, and by applying the laws of probability, they can project the possible number of solar mass stars that have already lived most or all of their active lives.

Current estimates are that there may be ten or fifteen billion white dwarfs in our galaxy alone. Some of these may be about the size of the Earth but yet have a mass equal to that of the Sun. A white dwarf is not actually the final stage in the evolution of a star; the final stage is that of a very cool world called a black dwarf.

White dwarfs have extremely strong magnetic fields and at this writing the most powerful magnetic field astronomers have ever been able to detect is related to a white dwarf in the constellation Sextans. This is relatively recent data and the white dwarf creating the magnetic field, PG 1031+234, was only discovered in 1986 by astronomers working with the 5-meter telescope at Palomar and the 2.3-meter telescope at Kitt Peak, as well as with data supplied by the International Ultraviolet Explorer Observatory satellite. PG 1031+234 (where 1031 is the right ascension for the dwarf and 234 is its declination), is a swiftly rotating star estimated to have a magnetic field 700 million times more powerful than the Earth's magnetic field. The dwarf is believed to have once been a normal star with a magnetic field that was greatly magnified when the star collapsed into its present state.

## QUASARS

Some of the mysteries of quasars, (quasi-stellar objects), were discussed earlier. These faint-blue distant members of the sky continue to confound astronomers at almost every turn of the telescope. They are, perhaps, the fastest things in the sky, many moving at 70 percent to 80 percent of the speed of light. Quasars may be as much as 10 billion light years away, yet they are as luminous as the light that may be collected from a hundred Milky Ways. Though they are so tremendously bright, they give off most of their energy in the infrared regions. Ultraviolet studies are expected to play an important role in unraveling at least one of their mysteries.

Exactly how many quasars are actually in existence is, like most things pertaining to numbers and distances in astronomy, a subject of a highly speculative nature. There may be a billion quasars in existence, but at the present time only a few thousand have been counted, and these few thousand show markedly different degrees of brightness.

Astronomers agree that the major sources of electromagnetic emissions come not from the entire breadth of a quasar but from very selected areas. Because of this, astronomers guess that the quasars may have a black hole at their center. This may be proof that they are actually very distant galaxies, although there is little hard evidence to confirm this idea.

Actually, astronomers are quite open-minded about whether quasars are galaxies or the nuclei of galaxies. They will be depending upon ultraviolet studies to help answer this puzzle. Ultraviolet instruments will be used to investigate the emission and absorption characteristics of quasar spectra. This will be accomplished by examining gas clouds in outergalactic and intergalactic regions lying between the expected location of the quasars and the point of observation. If it is determined that the clouds are with galactic boundaries, then quasars are surely not galaxies. If it is determined that quasars are outside galactic boundaries, then it is back to the drawing boards, for then quasars can be almost anything.

## NEBULAE

Nebulae are visible and invisible clouds of gas and dust floating between stars. The greater majority of the clouds cannot be seen even with the brightest telescopes, and those that are invisible can only be detected by special instruments. Nebulae can be separated into a number of different classifications but the simplest way to categorize them is to divide them into dark and bright nebulae. The bright nebulae are generally illuminated through the ionization of their gaseous elements or by the heat of nearby stars. The dark nebulae give off no light at all. But if the dark nebulae give off no light, how are astronomers able to detect their presence? Well, if the background environment of a dark object is lighter than it, then the dark object may be detected as a black blotch or patch in the foreground; and this is the way astronomers observe dark nebulae. The dark nebulae give themselves away whenever they pass in front of radiant clouds.

The solar system is suspected of having been formed from these great clouds, but there are other theories about how the solar system may have happened. Some theories suggest a solar collision that resulted in the scattering of material into space, these materials condensing into planetesimals which billions of years later formed the nine major planets and possibly the asteroids through the process of accretion.

Whether or not a great nebulas brought about the Sun and planets, astronomers remain curious about these great clouds. They seem to feel that they must play some role in the formation of stars. And if astronomers can be sure how stars form, they can begin to project about how the different types of galaxies have come to be formed. As the universe is a great system of galaxies, the more that can be learned about galaxies, the more astronomers can learn about the structure of the universe. Nebulae, therefore, are not merely of astronomical importance but cosmological and cosmogonical importance as well.

Many nebulae are highly complex systems, consisting of clouds within clouds and one or more stars populating the interiors. It usually necessitates extensive studies combining X-ray, infrared and ultraviolet techniques to unravel the complicated physics of a nebula. Take, for instance, the Orion Nebula, which is among the brightest of the great clouds floating in interstellar space. It inhabits a galactic area possibly five or six hundred parsecs from Earth and was for a long time believed to be one singular entity of interstellar gas. Infrared and ultraviolet studies, however, have revealed Orion's highly complex structure.

Orion is actually a visible cloud, though only barely so. As its name implies, it may be observed in the constellation Orion. It covers a group of stars—very hot O and B stars—collectively known as the Trapezium stars, the radiation from which is responsible for the fantastic glow of the nebula.[7] Orion is by no means one of the largest nebulae but it does have a radius that measures close to 3 parsecs. It is a complex system of hot and cold layers

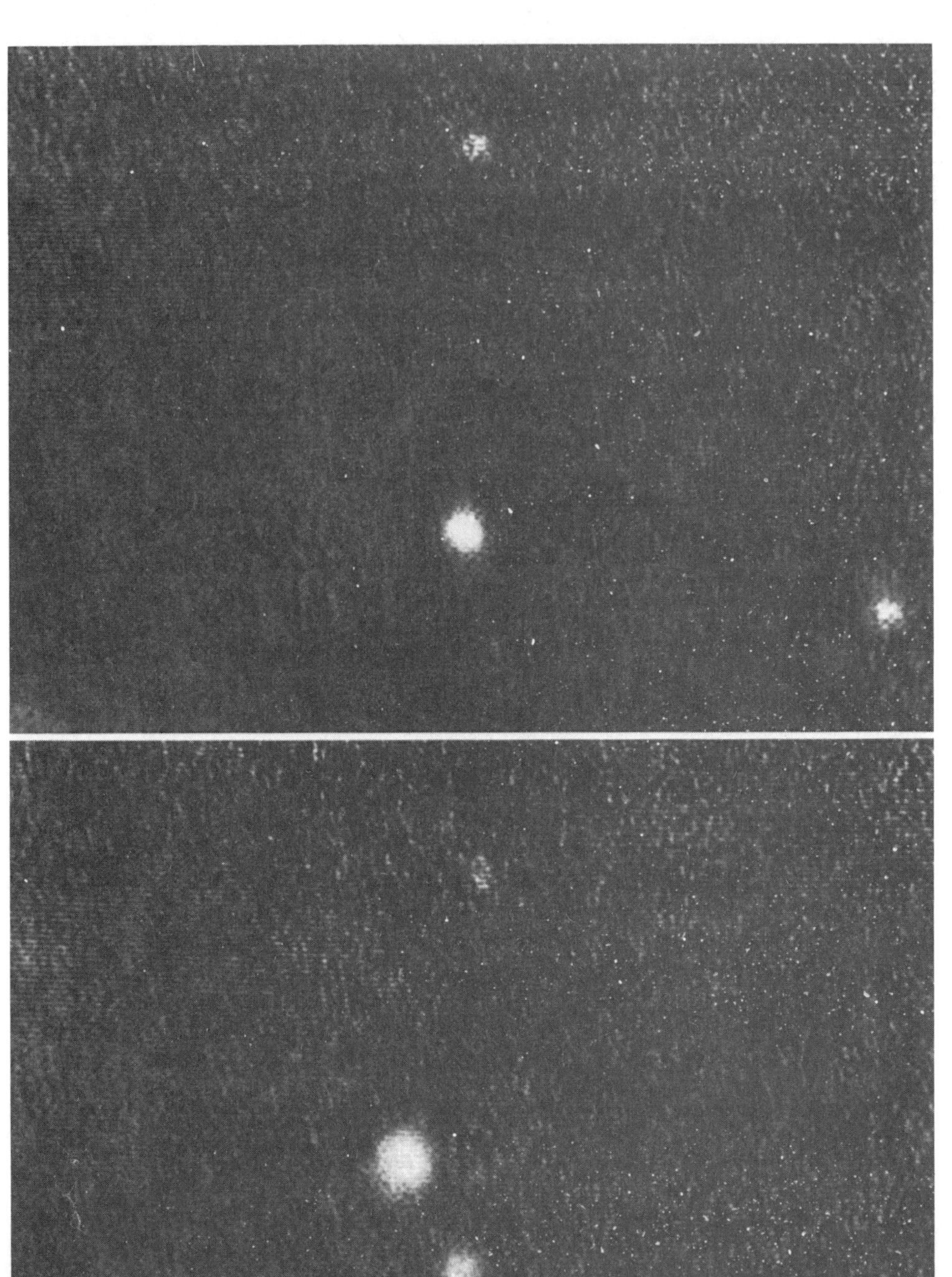

Fig. 8-8. Television photographs of the pulsar located in the Crab Nebula. The photos were taken at minimum (top) and maximum light. This pulsar may be a neutron star (Courtesy of Lick Observatory).

Fig. 8-9. The Orion Nebula (NGC 1976) and NGC 1977, showing faint outer parts in blue-violet light. Orion covers a group of stars collectively known as the Trapezium stars. It is the radiation from the O and B stars that make up the Trapezium group that gives Orion its fantastic glow (Courtesy of Lick Observatory).

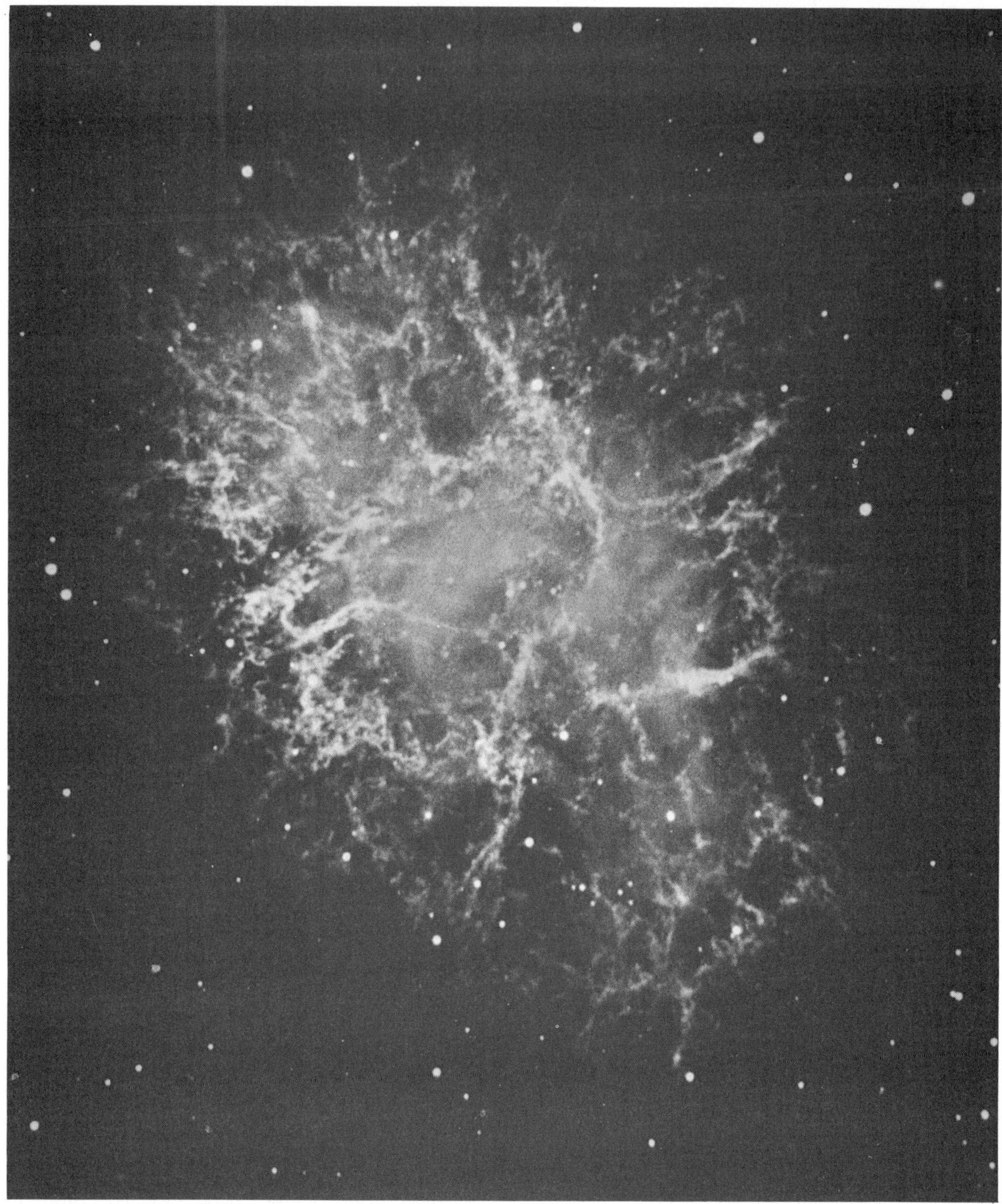

Fig. 8-10. The Crab Nebula, heralded as the most impressive cloud in interstellar space, is believed to be the result of a supernova explosion (Courtesy of Lick Observatory).

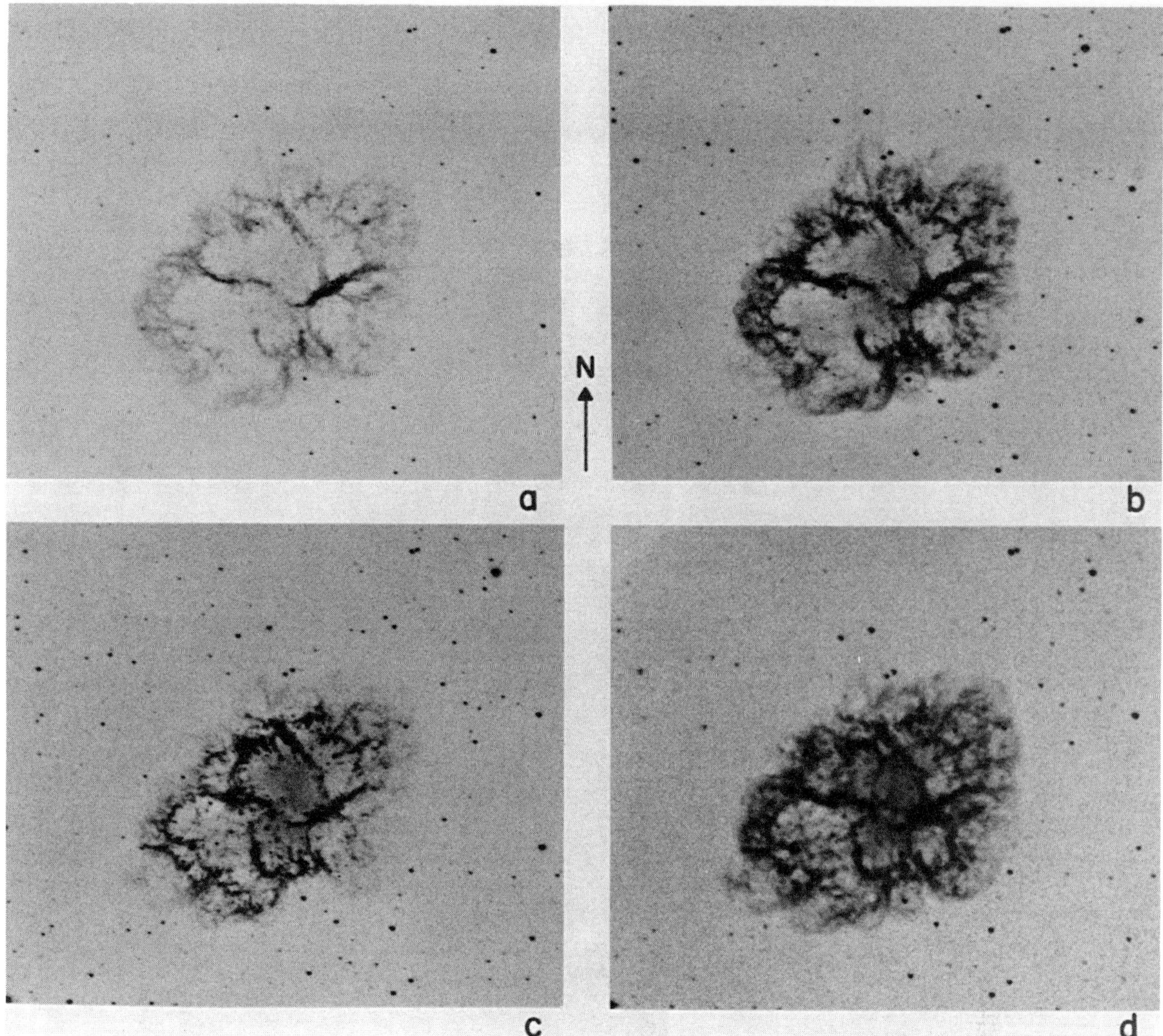


Fig. 8-11. These four photographs of the Crab Nebula were taken through various narrow-band filters. They show the detail in the expanding filaments. Photographs A and B illustrate doubly ionized gas; photograph C shows singly ionized sulphur; and photograph D shows singly ionized oxygen (Courtesy of National Optical Astronomy Observatories).

of clouds and stars, all surrounded by a neutral hydrogen shell that appears to be expanding. Eventually, the star clusters lying inside of it will burn away the cloud covering and the stars will burn brighter than ever. Right now, however, these are very young stars not quite a million years old, and their emergence from the cloud is probably not scheduled for millions of years.

By contrast, the Crab Nebula, noted by astronomers as the most impressive cloud in interstellar space, has as its core some pulsating object believed to be a neutron star. The cloud itself is believed to be the result of a supernova explosion that occurred in the sixteenth century. It is roughly 3 × 2 parsecs and gives off very strong emissions in the infrared, ultraviolet, X-ray, and gamma ray regions of the spectrum. With a mass about equal to that of the Sun, the cloud is believed to consist only of

Fig. 8-12. Pierre Simon Laplace (1749-1827) was the French astronomer and mathematician who believed the solar system formed from some great nebula that began to contract into a series of "rings" which served as the fundamental material for the formation of the Sun and planets (Courtesy of Yerkes Observatory).

the remnants of the supernova; nothing else has become attached or absorbed into it. There remains to be a marriage between the supernova's material and that debris which ordinarily lies in or orbits through interstellar regions.

Ultraviolet research has detected the presence of hydrogen molecules in nebulae, and this has resulted in a rather significant contribution to cosmological models. Using statistical data based on the relationship between hydrogen molecules and atoms with other cosmic components, as well as using complex formulas to feed into their computers, astronomers have been able to determine that there may possibly be enough hydrogen molecules in the universe to allow the formation of two or three billion stars with up to six times the mass of the Sun.

## DESTINY

Ultraviolet astronomy will play its role in helping astrophysicists determine the ultimate destiny of the universe. One of the great questions still needing an answer is whether or not our expanding universe will continue to expand forever or else will one day come to a halt and then begin contracting. Once the contraction phase begins, ultimate doom awaits everything which exists.

The answer to the future of the universe depends upon the relationship between mass and gravity. That is, astronomers can find the answer to this great question by measuring the total mass of the universe and then determining to what extent gravitational forces will act upon that mass. If the mass/gravity relationship is such that the expansion phase can be slowed, then eventually the galaxies will no longer continue to recede. If this is indeed the case, then cosmologists predict that for the next 60 billion years, the current expansion phase will continue and at probably the current rate of expansion. Then the rate of expansion will begin to decrease for another 50 billion years or so until it finally slows to a complete halt and a phase of contraction begins. During the contraction phase, the universe will gradually, then quickly, be reduced to the singular and simple body that it must have been at its birthtime.

Most astronomers and astrophysicists, however, expect that the universe is an "open" rather than a "closed" system. They fully expect it to expand for hundreds of billions of years. But galaxies cannot continue ad infinitum because the stars which populate them must eventually burn themselves away, and as they die away so will the galaxies. Just as other stars must necessarily die, so too must the Sun. It will burn away, and as it does so—before it does so—Earth and the other major planets will have been obliviated. But this does not mean total destruction. New stars are always being born and probably new galaxies will form. The stars and planets we know today may be gone in the next 100 billion years, but there will be new ones to take their place. And as for the human race, if it has not blown itself away by then, in 100 billion years it may be on another or other planets, in another galaxy, or in another universe.

At the present time, research reveals that there is not enough matter existing in the universe for gravitational influence to halt the expansion phase, but astronomers cannot be sure. It is impossible to total up the mass in any cosmic field and project what might actually be the total mass in the universe, because it appears that up to 90 percent of the mass in the universe may be missing. In other words, astronomers expect that there should be more matter existing in the clusters of galaxies than they are able to detect. They know the matter must exist because they are able to detect its gravitational effects. For instance, there are clusters of galaxies in which the members appear to be moving much too fast to remain joined in the clusters by their gravitational interactions. These galaxies can only remain a part of the clusters if there is anywhere from 10 to 90 times the amount of detectable matter within the clusters. This matter may be contained in dark halos observed to be surrounding certain spiral galaxies, but astronomers cannot be sure that these dark halos contain the same type of matter that should be found in the galactic clusters or throughout any selected spacial field.

The question of the missing mass is one of the major problems confronting astronomers and cosmologists. Actually, some researchers feel that the term "missing mass" is somewhat misleading, for they believe that the mass is by no means missing but is, rather, just undetected. They would argue that the term "hidden mass" is much more appropriate. Others, however, feel that the amount of mass theorized to be necessary for the current order of things is indeed missing and it's up to science to find out why the universe has no need for it. As to what the hidden mass may actually be, there have been many speculations, including the existences of hypothetical particles and non-reflecting matter.

How will ultraviolet research help solve the great question of the hidden or missing mass? Astronomers have based their theories about the quantity of matter in the universe being much too low to inhibit the expansion on the amounts of helium and deuterium they can detect in space. Helium is the second most common element in the universe and is believed to have been formed during the first few minutes after the great explosion that resulted in the birth and/or expansion of the universe. But helium is also created during stellar nuclear fusion reactions, in which case it owes its existence to hydrogen, from which it is synthesized. Hydrogen is the element which is most common throughout the known part of the universe and deuterium (heavy hydrogen) is a hydrogen isotope having twice the mass of ordinary hydrogen. As the cosmological theories go, the abundance of deuterium existing at the moments after the Big Bang would have depended upon the density of matter at the time. So, astronomers theorize that if they examine the helium and deuterium contained in intergalactic clouds, they can deduct what was the true relationship between helium and deuterium when the universe was first forming, and can determine if the hidden mass is actually there or poor theory—like ether was until the late nineteenth century. And it is through ultraviolet research that astronomers will be able to detect and measure helium and deuterium abundances in nebulae.

# Chapter 9
# X-Ray Astronomy

IN NOVEMBER OF 1895, A GERMAN PROFESSOR OF physics by the name of Wilhelm Röntgen was experimenting with the way in which electric current flowed in a cathode ray tube. Little did he realize that his research would lead to a Nobel Prize in Physics (1901), for what he was to uncover was by no means the objective of his search, and the experiment itself was only one of many that he conducted regularly. The results of his efforts would launch a brand new era of physics—indeed, a brand new physics—and thrust medical research onto a new plateau.

As Röntgen carefully observed and recorded the events of his experiment, he began to notice that there was a direct relationship with what was occurring in the cathode ray tube and what was happening to the fluorescence of a nearby chemical substance. He made a wild guess that some new type of radiation was resulting from the behavior of the electrons in the tube. Later, he noticed that this new form of radiation seemed to penetrate many different kinds of material. He was not exactly sure of the kind of radiation on which he had stumbled, but of this he was certain: this radiation, thereafter called Röntgen radiation, probably was not in anyway related to the type of radiation that produces light, for Röntgen radiation seemed not to have the reflective, refractive, interference, and polarization properties of light. Yet, the rays did propagate in straight lines, were not deflected by magnetic or electric influences, and were not absorbed by matter in proportion to the density of the absorbing material.

The scientific community reacted with great enthusiasm to Röntgen's discovery. All sorts of suggestions, papers, and arguments were produced to describe what might be the physical nature of this new radiation. Some physicists guessed that Röntgen's rays were acoustical or gravitational waves at extremely high frequencies. By 1900, the German scientific community was beginning to argue that the radiation was no more than ordinary light rays at the higher frequencies. Röntgen himself hoped that his new discovery could be related in some way to the very controversial ether wave, but there was no convincing his colleagues of this. Almost from the beginning of his discovery, physicists believed that the radiation was a part of the electromagnetic spectrum.[1]

## X-RAY OBSERVATIONS

Röntgen radiation is also known as x-radiation; actually, Röntgen had named the radiation x-radiation but his colleagues immediately dubbed it Rontgen radiation. Röntgen was wrong in some of his first assumptions about these new rays. X-radiation is in no way related to the illusive ether and is definitely the same kind of radiation that produces visible light; it is simply electromagnetic radiation at wavelengths from 12 nanometers to 0.0024 nanometers. The X-ray region, therefore, is the region that lies between the XUV and gamma ray regions of the spectrum. X-rays have come to be conveniently divided into two main groups, low energy (soft) and high energy (hard).

Like radio, infrared and ultraviolet astronomy, X-ray astronomy is important in the study of almost every type of celestial object, from the Sun to the very distant quasars. It is a particularly important science for studying those areas in space containing stellar or galactic systems which appear to be the scene of mysterious and certainly violent events resulting in high-energy particle or gas emissions. There are, of course, certain problems related to X-ray surveys of the sky and these have to do with the terrestrial atmosphere which serves as a blockade, so to speak, against raiding X-rays. Thus, X-ray astronomers must depend upon high altitude balloons or rockets, the balloons serving to detect the hard X-rays and the rockets serving to detect the soft rays. There are still observational gaps in the lower energy fields because a lot of the emissions coming toward the Earth are absorbed by hydrogen gas clouds drifting in interstellar regions.

X-ray telescopes are somewhat different from those used in the other astronomies. They are either of the focusing or non-focusing type. In the focusing type, a rather unusual optical design is employed, necessary because of the peculiar properties of X-rays. X-rays behave just like visual light rays and they bounce off objects for exactly the same physical reasons. Because of the size of the energies involved in X-ray emissions, the refractive index (ratio of the speed of radiation in any two media) is not the same as that for visual light, and reflection only occurs under certain conditions. These conditions depend upon wavelength and the type of surface being bombarded. Lower wavelengths (and, therefore, higher energies) require small grazing angles. X-rays are at the lower wavelengths of the spectrum so reflection is only high in those cases collected rays only graze the surface of telescope mirrors. The amount of actual reflection varies in proportion to the various wavelengths at which the X-rays are collected, as well as by the X-ray absorption edges in the material used for reflecting the rays. The problem of collecting these rays has been solved in a number of ways, one of which has been through the design and manufacture of nickel-coated mirrors shaped especially for the reflection of X-rays. These mirrors are placed inside a telescope so as to produce two-dimensional images for study.

In the nonfocusing type of X-ray telescope, proportional counters or scintillation detectors are included as part of the design to maintain directional sensitivity.[2] A proportional counter is an electronic device filled with gas that becomes ionized whenever x-radiation or other ionizing radiation reaches the gas. The counter sends an output pulse to indicate the number of electron-ion pairs that have resulted from the ionization, thereby also measuring the amount of radiation collected. Scintillating detectors are more useful for gamma ray detection but may be used for x-radiation detection. These detectors, or counters, are made up of scintillating crystals that flash whenever they sense radiation. A pulse is produced by a photomultiplier to correspond to the activity or energy level of the incoming radiation. In certain observatories, and space missions, imaging proportional counters as well as gas scintillation proportional counters are also in use.[3] The first expands the capability of the standard proportional counter by creating a two-dimensional image; the second doubles the energy resolution available from standard counters.

## SOLAR X-RAYS

The practical beginnings of X-ray astronomy

may be traced to September 1949 when a Naval research laboratory team, headed by Herbert Friedman, launched a V-2 rocket carrying instruments designed to detect both extreme ultraviolet radiation and x-radiation. The extreme ultraviolet detectors were the first to signal back success, the first signals actually coming as soon as the rocket was about 75 kilometers above the Earth. Another ten kilometers into the sky, and the X-ray detectors began signaling. Other instruments told Friedman and his group that the X-ray emissions were coming from the Sun. The astronomical community had long suspected that the Sun was a source of x-radiation but until this time, they could not prove it. In 1950, subsequent studies verified that the solar X-rays were coming from regions associated with sunspots.

Sunspots are dark blotches or streaks occurring in photospheric granulations. These "spots" can be hundreds of millions of kilometers in area but with the gigantic Sun as a backdrop, they do indeed appear as little more than spots (Fig. 9-1). Sunspots have two distinct regions, the umbra and penumbra. The umbra is the dark core of the spot and the penumbra is the lighter area surrounding the core. Sunspots come and go. Generally they remain visible for periods of two to three weeks, but they can continue to exist for periods of two to three or many months. They tend to occur in clusters and in the environment of strong magnetic fields. The number of sunspots fluctuates over a period running from about 9 to 14 years; this period is generally referred to as the *sunspot cycle* or the *solar cycle*.

Among the phenomena associated with the sunspot cycle are tremendous eruptions of intensely hot gas. These are the spectacular solar flares in which astronomers are becoming more and more interested (Fig. 9-2). They are of particular interest because the flares create tremendous pressure upon the solar wind which, in turn, moves toward Earth with such force that it causes all sorts of problems, some of which can be particularly dangerous. The lesser problems have to do with electromagnetic disturbances; the greater problems have to do with maintaining and protecting orbiting spacecraft. Thus, understanding flare phenomena is extremely important, but to the time of this writing, there is very little astronomers and astrophysicists know about flares. Here is where X-ray astronomy becomes so very important, because it is one of the major means by which astronomers can gather information about these dangerous solar explosions.

Flares vary considerably in both intensity and form but the tremendous energy related to the more violent flares is underscored by the fact that within half an hour after an explosion, the Earth is already being bombarded with cosmic rays or lesser-charged particles. The latter are the cause of sudden changes in the Earth's magnetic field; these changes and disturbances to the magnetic field are called geomagnetic storms. The initiation and development of these forms follow no noticeable pattern, but in some cases, four phases of development can be associated with the phenomenon.[4] These are:

- Sudden Commencement Phase, which occurs as a result of a shock wave, produced by a solar flare, and which causes an increase in field strength through compression of the magnetosphere. This rise in field strength generally occurs over a 2 to 5 minute period.
- Initial Phase, which occurs as the result of the Earth being surrounded by the postshock plasma and field. This phase may possibly last for many hours, during which time the surface field strength is at levels in excess of those occurring during the commencement phase.
- Main Phase, which occurs as the result of increasing particle population or acceleration and is marked by a decrease in the surface field strength for a period of up to 24 hours.
- Recovery Phase, which occurs as the result of the decaying ring current, and is marked by a return to or near the sudden commencement phase field strength.

A knowledge of solar flares not only aids astronomers in predicting and guarding against the dangers of solar flares but also helps them understand something about flare stars, which are red main-sequence stars that suddenly increase their lu-

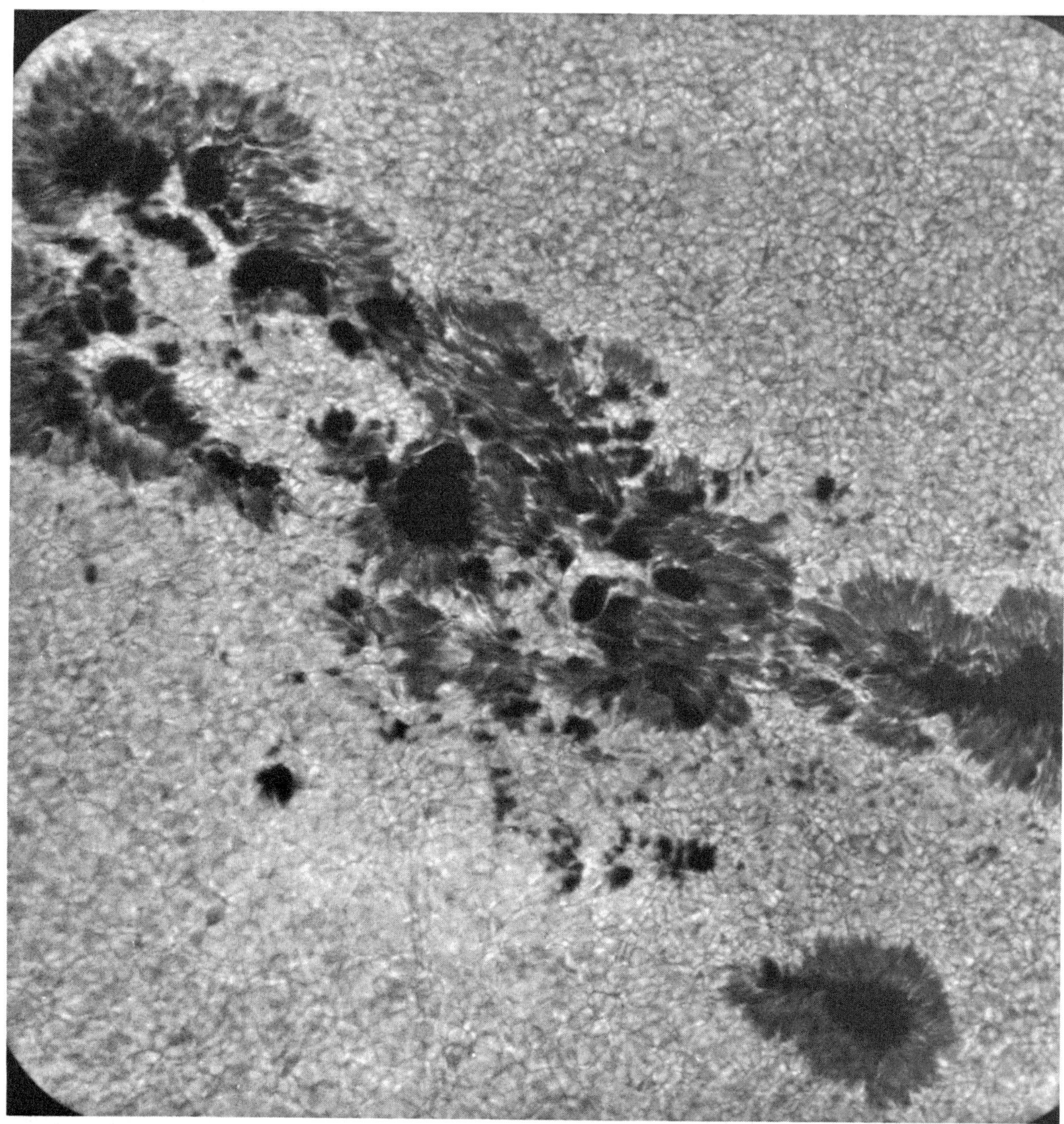

Fig. 9-1. A white light photograph of a complex sunspot group and granulation of the Sun. Sunspots can be hundreds of millions of kilometers in area. They consist of a dark core called the umbra and a lighter area called the penumbra (Courtesy of National Optical Astronomy Observatories).

minosity. The sudden brightness change in these flare stars does not occur throughout the stars but only in very limited regions. The changes in brightness also occur over periods of relatively short time—less than a half hour or so—during which time the peak of luminosity is reached and a return to original brightness occurs. These flares can be detected not only at optical wavelengths but also at radio and X-ray wavelengths. In many cases the emissions are not coincident and radio or X-ray flares are produced independent of each other or independently of the optical flares. These flare stars are generally Me stars with magnetic fields probably equal to that of solar strength, thus one of the reasons why astronomers hope a knowledge of solar flares will lead to the compilation of greater data about flare stars.

It is still a mystery about what causes flares. Perhaps, they are the result of internal interactions or perhaps they are the result of reactions occurring in the solar atmosphere. If there were some way that astrophysicists could duplicate solar flare activity in the laboratory, they might be well on their way to finding the answers to the mysteries of the flares. But laboratory duplication is impossible, so a lot depends on electromagnetic readings, particularly in the shorter wavelengths. Flares are believed to be up to 10 million degrees in temperature and, therefore, extreme ultraviolet, gamma ray and X-ray studies are the key to learning more about them.

Another of the major problems confronting astronomers is the fact that there is no way they can analyze a flare in complete detail using one single type of radiation collecting instrument. Flares may

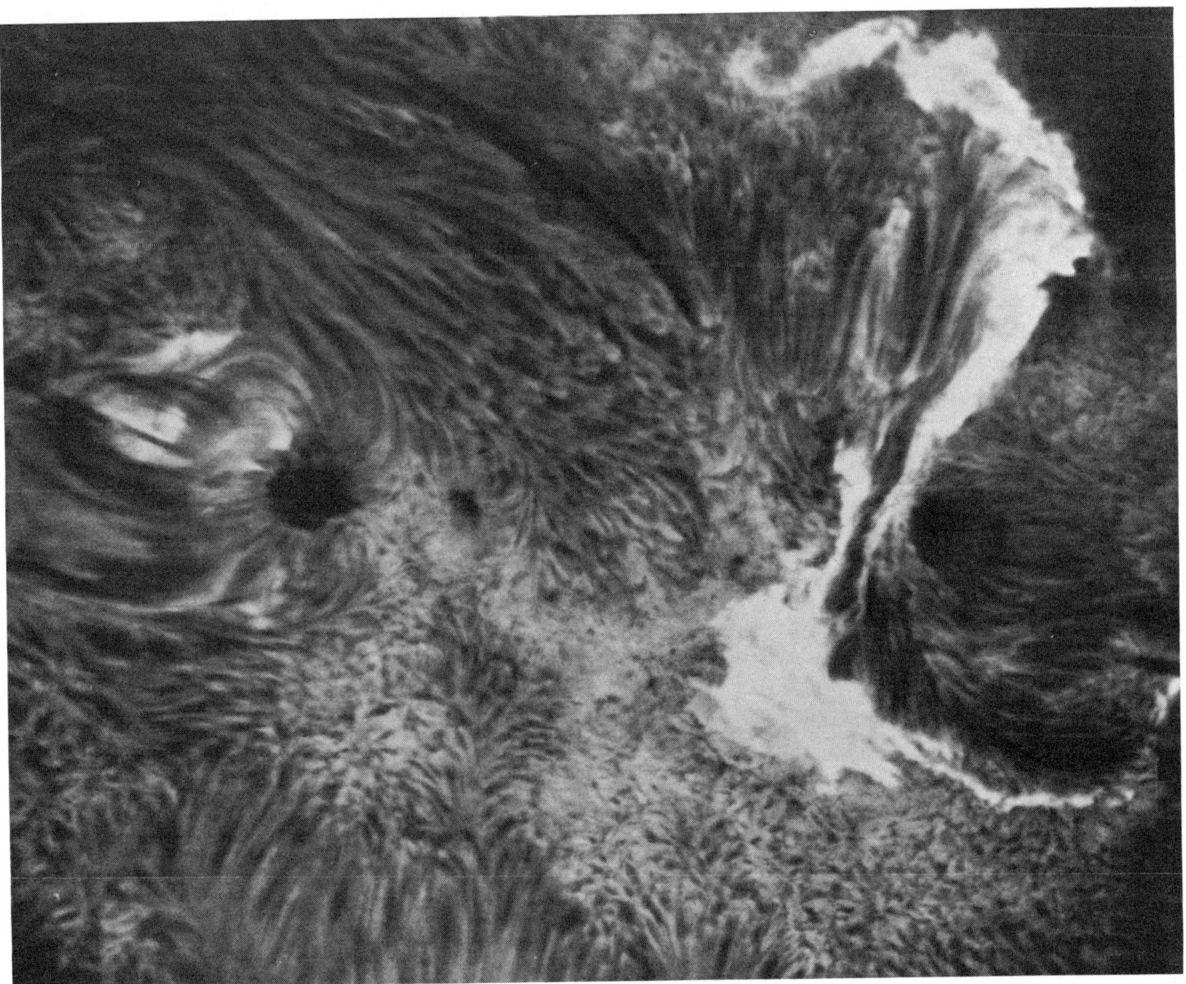

Fig. 9-2. Solar flares create tremendous pressure on the solar wind, which can be particularly dangerous when it blows as far as the Earth's atmosphere. These solar flares have been a tremendous mystery in the past but with the growth of X-ray astronomy many of these mysteries may soon be solved (Courtesy of National Optical Astronomy Observatories).

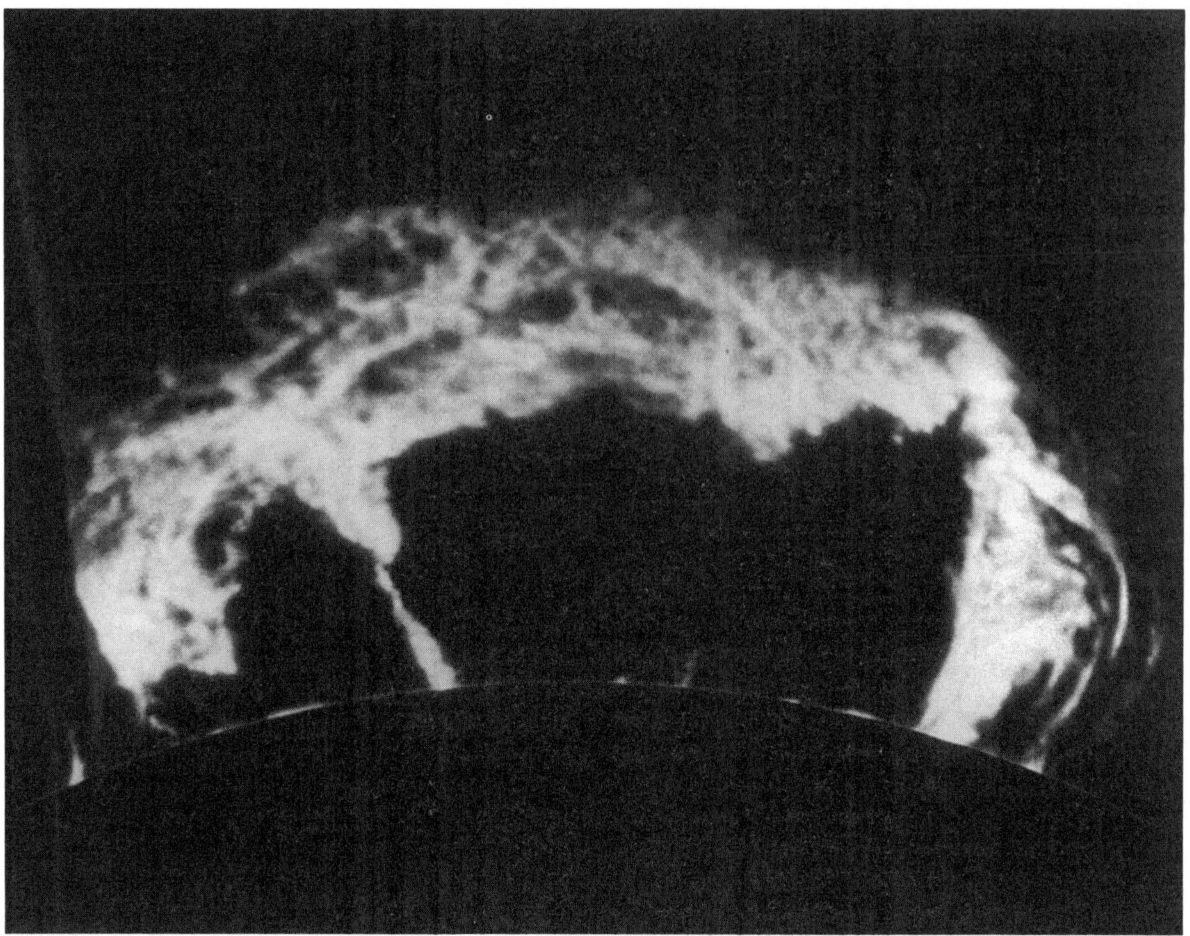

Fig. 9-3. A prominence erupts from the solar surface. Prominences are clouds of gas of much higher density but lower temperature than their immediate environment. They are called filaments when they are viewed against the bright disc of the Sun and appear as dark absorption features (Courtesy of National Optical Astronomy Observatories).

reach temperatures of 10 million degrees but they generally have a wide range of temperatures throughout their regions and in some regions that may be as "cool" as 10 thousand degrees. Different temperatures produce radiation at different wavelengths, so an extensive range of equipment is necessary to help put the pieces of the flare puzzle together.

X-ray studies depict flares as loop-like flames spreading over thousands of miles and reaching out from the Sun for hundreds or thousands of miles. Still, they are little more than tiny specks to the astronomer, for when we are talking about solar measures, hundreds of thousands of miles represent only a small portion of the total area being investigated. Astronomers theorize that when these loops collide, they produce the fantastic explosions that hurl solar particles to Earth in less than a day.[5] Highly charged particles race through the loops and interact with the flare plasma in such a way that radiation is produced.[6]

X-ray studies have been particularly useful in helping astronomers analyze the characteristics of solar flares because the flares are very rich sources of x-radiation resulting from extremely rapid heating and particle acceleration. Once astronomers collect solar X-rays, they can determine a great deal about internal temperatures and may one day be able to say with certainty what is happening within the flare and just what is the cause of its frequent and violent eruptions.

More specifically, X-ray studies of the Sun have:[7]

- Aided in determining the temperatures of flares
- Presented a detailed history of flare phenomenon
- Revealed flare plasma to be complex states with significant temperature changes over short intervals
- Discovered coronal flares, which appear within hours of a major flare
- Revealed a great X-ray arch over the area of decayed flares, evidently related to postflare radio storms
- Assisted in developing flare maps of the Sun, important for eventually learning why some sunspot groups are explosive fields and some are not

Technically, although the Sun emits a great deal of x-radiation and is indeed a star, it does not qualify

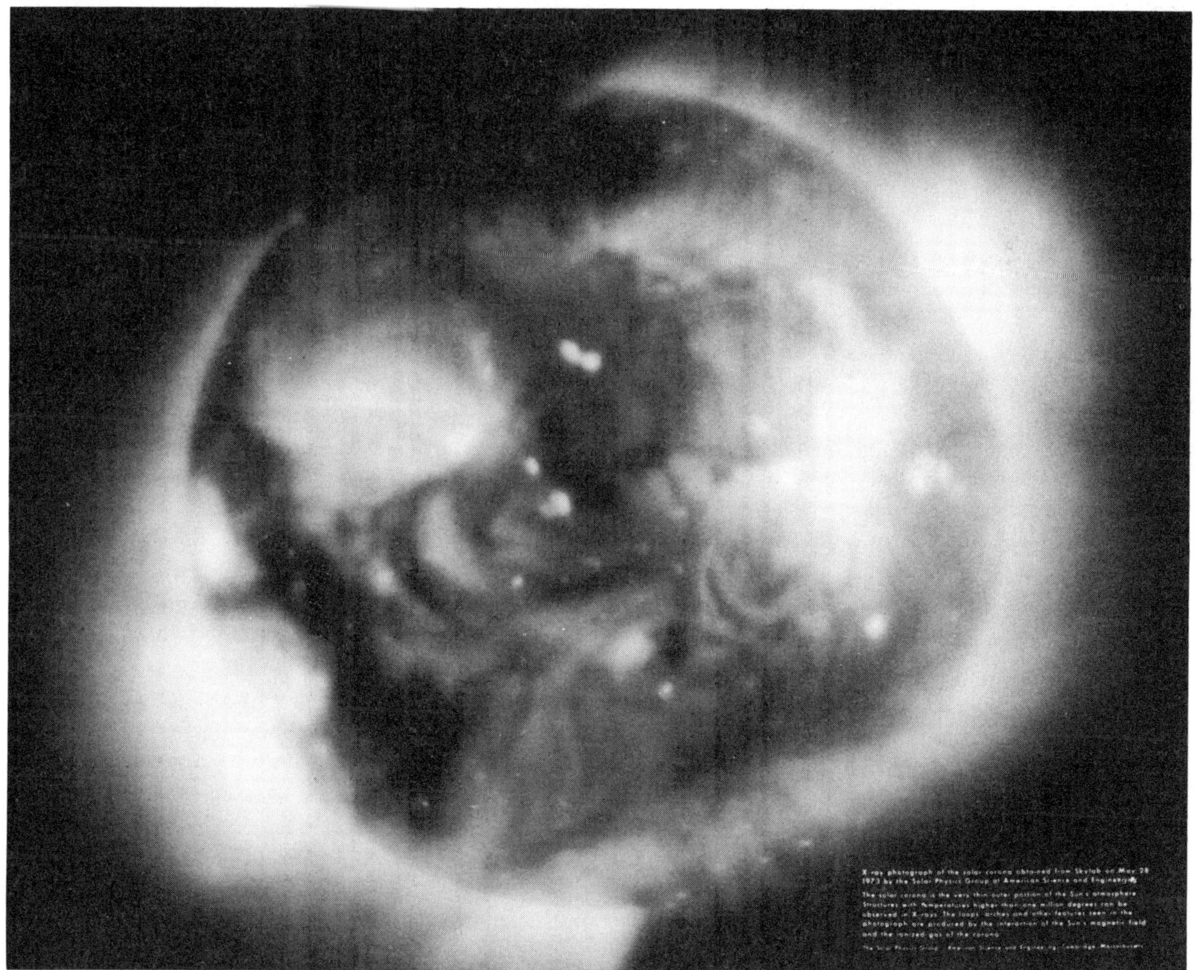

Fig. 9-4. The solar corona in X-ray. By eliminating all other radiation and accepting only X-rays, astronomers can obtain an image of the hottest parts of the solar atmosphere. The outer layer of the Sun, the corona, has temperatures which probably exceed one million degrees (Courtesy of NASA).

as an X-ray star because its optical radiation is far more intensive than its x-radiation.

The United States and Japan will soon be joining forces to design and operate a soft X-ray telescope. The telescope will be used in SOLAR-A, the Japanese solar observation satellite, scheduled for launching in 1991. In addition to the joint effort on this soft X-ray project, the Japanese will be developing a hard X-ray telescope for the study of flares at temperatures in excess of 30 million degrees Kelvin. The Japanese and American teams expect that by employing both telescopes for the study of solar flares, they will be able to achieve a greater knowledge of the solar flare cycle.

The Sun is only one of many sources of X-rays in vast space. Many more sources exist beyond our solar system but within the Galaxy; these are generally referred to as X-ray binaries. Still more sources of X-rays lie well beyond the Galaxy and these include:

- Galactic clusters
- Radio galaxies such as Virgo A, Centaurus A and 3C 236
- Quasars

## X-RAY STARS

X-ray sources lying beyond the solar system but with the Galaxy are generally defined as X-ray stars, although in some cases the true source of the radiation may be black holes or some type of interstellar material. As far as can be determined, just about all X-ray stars occupy a narrow field near the galactic plane or else regions within the galactic halo.

The first X-ray star was detected in 1962 during a joint experiment by teams from American Science and Engineering, a private company based in Cambridge, Massachusetts, and the Massachusetts Institute of Technology. Leading the experiment was a young physicist by the name of Riccardo Giacconi, who, by the age of 28, had already acquired some impressive credentials. He had been a lecturer at the University of Milan where he engineered the largest multiplate cloud chamber in Italy, assisted in the development of scintillation chambers through the use of image intensifiers, had been involved in many phases of cosmic ray research, and was now ready to redirect his interests to space science.

The 1962 experiment was not the first attempted by Giacconi. He had been involved in related experiments in 1960 and 1961. The 1960 effort terminated when the Nike-Asp rocket engine his team was working with failed; and the 1961 experiment ended when the doors to another rocket failed to open and, therefore, to transmit data back to them.[8] In June of 1962, however, the team met with success when their rocket, equipped with three proportional counters no larger than credit cards, began signaling back the presence of a strong X-ray source somewhere in the southern sky.[9] At first, Giacconi and his team were highly skeptical because the signals were much too strong. There was little possibility that anything beyond the solar system could be emitting x-radiation with such strength. Later analyses of the data transmitted from the rocket proved the source was indeed outside the solar system and specifically within the constellation Scorpius.[10] Quite appropriately, the star was named Sco X-1, short for "the number 1 X-ray source 1 in Scorpius."

The experiment that led to the detection of Sco X-1 also produced evidence of a uniform distribution of x-radiation throughout space. This X-ray

**Table 9-1. X-Ray Astronomy.**

| System | Major Sources |
|---|---|
| Galactic | X-ray binaries, supernova remnants, some nebulae, neutron stars, black holes |
| Extragalactic | Seyfert galaxies, radio galaxies, quasars, clusters of galaxies, BL Lac Objects |

background radiation serves as the backdrop for all X-ray observations of galaxies and X-ray stars; all X-ray observations must take into account this existing background radiation in order for studies to be successful. In optical astronomy, visual light rays representing planets, stars or galaxies rest upon a darkened field, and so they are easy to spot. In X-ray astronomy, however, stars and galaxies do not appear in the mostly black and white theater we are used to at optical wavelengths; instead, these objects are set against a glowing background, equally bright in all areas. No one knows the reason for this cosmic glow of x-radiation but astronomers suspect that it is being emitted from clusters and other galaxies not yet identified.

Since its discovery in June 1962, Sco X-1 has also been detected optically. Like most X-ray stars, Sco X-1 is a member of a binary star system, this one lying about 500 parsecs from Earth. The constellation in which it resides is extremely bright, contains many star clusters, and is highly visible to the unaided eye. The brightest star in the constellation is Antares, an advanced star that may be up to 500 times the size of the Sun, The star, however, with which Sco X-1 has been optically identified is V818 Sco, a variable star whose brightness is measured at the 13th magnitude and, therefore, not observable by the unaided eye.

Most X-ray stars are believed to be stars well into their evolution; probably three-quarters of their lives have already been spent. Why and how they emit X-rays, however, are questions which astronomers still have to answer. The general idea is that they are continually receiving hot, gaseous materials from their companion stars, and it is this gas which is believed to produce the strong X-ray emissions.

It is quite possible that these X-ray stars are those white dwarfs which represent stars of relatively low mass and which have entered into the final stages of their evolution. These stars, you will remember, have exhausted most of their nuclear energy and have collapsed into worlds having extremely small diameters but extremely high densities. Eventually, these white dwarfs collapse even further and become neutron stars or black holes. Those white dwarfs that become neutron stars, as current theory goes, will be the source of those quick but even bursts of radiation that give the star the same pulsating features (at X-ray wavelengths as well as radio) that are ordinarily detected from pulsars. Those that become black holes continue to emit x-radiation in long and steady streams.

Both neutron stars and black holes are subjects of continued and creative speculation in astrophysics.

## NEUTRON STARS

Neutron stars can also come about as the result of supernova explosions, which result from the gravitational collapse of the star. The collapse occurs because the star's own gravitational forces cause it to enter into a contraction phase of relatively short duration. The star becomes so heavy that the pressure produced by its internal gases can no longer support its construction. The result is a massive explosion that blows most of the star's debris out into space, converting the remainder of the star into a compact world of high density or converting it into a massive gas cloud.

Astronomers seem to be relatively sure that the radius of a neutron star will be approximately 10-12 kilometers and that its mass will be, on the average, about 1.5 times the mass of the Sun. Because some neutron stars are part of binary systems, it is relatively easy to determine their mass by measuring the gravitational interactions within the two-star systems. If the neutron star should obtain a mass greater by 5 times than the Sun's mass, it would collapse into a black hole. These determinations of neutron star mass are based on foundations more theoretical than factual, and it is certainly quite possible that neutron stars can actually be many more times as massive as the Sun.

Neutron stars are expected to have magnetized surfaces of gas, becoming solid at depths of a few meters. The interiors of these objects are made up mostly of neutrons, which are baryons expected to be present in all atomic nuclei except hydrogen. (Because it has a zero charge, the neutron has very little electromagnetic effect on other subatomic particles and it easily penetrates an atom). Inside

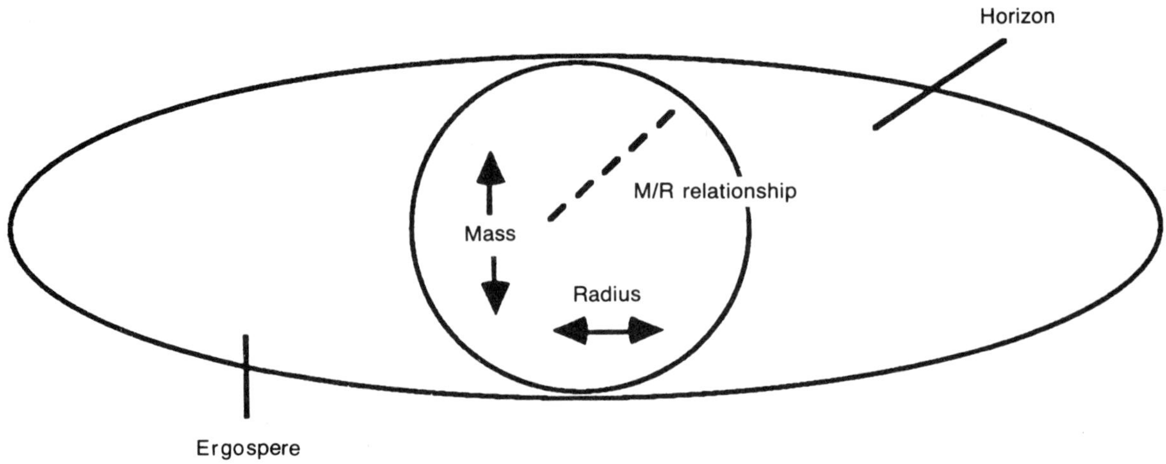


Fig. 9-5. One of the possible results of the gravitational collapse of a star more massive than the Sun is the creation of a black hole. There is no direct evidence of their existence, but x-ray detections of matter hurtling to a place of no return help give support to arguments of their existence.

the neutron star is a deep sea of superfluid which flows between the solid crust and a possibly solid core. The matter inside the star is in an extremely dense form and the physical laws governing the subatomic constituents in this superfluid state result in something called *degeneracy pressure*. Pressure is usually temperature dependent, but in the case of degeneracy pressure, it is density dependent. Because of the very tight packing of the neutron particles which cause the degeneracy pressure, the neutron star maintains its stability regardless of the range of internal temperatures.

Little more can be ascertained about neutron stars. Hopefully, X-ray studies will unravel many of the mysteries related to them, particularly those mysteries about what is actually occurring deep inside the neutron stars and what particles may be involved in internal interactions. Right now X-ray astronomers are running into more mysteries than they are solving. For instance, while in many cases they are able to detect evidence of core stars in supernova remnants, in many other cases there appears to be no evidence of a neutron core.

## BLACK HOLES

Black holes always seem to capture everyone's imagination. They are discussed, written about, and illustrated so frequently that everyone takes their existence for granted. The truth of the matter is that they are hypothetical existences, just like the quark.

Fundamentally, a black hole is a theoretical object in a very advanced state of collapse. It is well beyond the neutron star phase. Its gravitational pull is like nothing else in the universe. Anything which comes into the black hole's gravitational field and is moving less than the speed of light will be captured, and there is no escape.

As theory has it, a black hole is what remains of a star that was once five to ten times more massive than the Sun, but has since gone through a series of nuclear reactions that culminated in a catastrophic explosion. During the explosion, the outer shell of the star and all of the gas and lighter elements were blown away, and the heavier elements in the star's core immediately began to collapse.[11] The scenario here is much the same as that for the making of a neutron star, but in the case of the black hole, the collapse continues until all the material in the star is squeezed into what may actually be an invisible point in space.

This mere point in space, once a star as great or greater than the Sun, is a brand new cosmic entity in that it no longer contains any of the properties of the star from which it was produced. Like a subatomic particle, it can now only be described in terms of its mass and spin.[12] But if its gravitational field is so strong that nothing--not even light—can escape from it, how can it be detected? Most stars are binaries, orbiting about each other. They are, in a way, satellites of each other; one star goes about the other as the Moon goes about the Earth. In this case, the black hole will have had a living companion (a true star). As would be expected, debris from the living star will be drawn into the gravitational field of the black hole. The material from the living star will produce x-radiation as it is sucked into the black hole. The physics of the situation goes something like this: As loose matter gets caught in the gravitational field of the black hole, its speed increases. The closer it is drawn to the source of the gravitation, the more the matter becomes subject to the stronger and stronger gravitational interaction. The stronger the gravitational field, the faster the matter races toward . . . oblivion. As the speed of the matter increases, so does the amount of friction produced. The friction creates intense heat, allowing for temperatures up to, perhaps, hundreds of millions of degrees. At these extreme temperatures high energy radiation is produced and so gamma ray and X-ray emissions can be detected. Thus, the existence of black holes may be detected not by their emissions but rather by the emissions of the matter being drawn into them.

The strongest evidence of the existence of black holes comes from a binary X-ray source called Cygnus X-1. This is an extremely powerful galactic source of x-radiation that astronomers associate with an extremely bright supergiant identified as HDE 226868. Studies of optical data have indicated that the X-ray source associated with the supergiant may be in the range of 10 solar masses. This would qualify the X-ray source as either a neutron star or a white dwarf—or possibly a black hole, if indeed a black hole represents the next stage of evolution beyond the dwarf or neutron star stage.

X-radiation from Cygnus X-1 is being detected in quick but irregular patterns, indicating that the emissions are caused by the pull of gas from the supergiant to the X-ray source, much as astronomers would expect to happen if a black hole were part of the system.

At this writing there are only three known stellar black holes, Cygnus X-1, LMC X-3, and A0620-00. The last is a very recent find (1986) by Jeffrey McClintock of the Harvard-Smithsonian Center for Astrophysics and Ronald Remillard of MIT. A0620-00 is by far the closest black hole to Earth, possibly only 1,000 parsecs from Earth.

There are other very strong sources of x-radiation from within the Galaxy, among them Centaurus X-3 and Hercules X-1. Centaurus X-3 was the first X-ray binary to be discovered, as a result of the Uhuru satellite experiments in 1971. Hercules, the second X-ray binary to be discovered but the first to be optically identified, emits a complex pattern of radiation. Neither of these two sources are suspected of being black hole systems. Rather, they are expected to be neutron stars.[13]

A new black hole candidate has been observed in galaxy M-82, located in the constellation Ursa Major. The X-ray emissions from this candidate, according to John Stocke of the University of Colorado at Boulder, are stronger than any so far detected.

## EXTRAGALACTIC SOURCES

Beyond the galaxy are countless sources of x-radiation. In many instances, these extragalactic sources are actually radiating more strongly at the X-ray wavelengths than at the optical. The major source of these X-rays are both normal and active galaxies, galactic clusters, and pulsars.

Emissions from normal galaxies do not always appear to be coming from a central source but rather from the stellar population in general. As an example of a normal galaxy with diffuse X-ray emissions, we might consider another member of the Local Group, Andromeda, which is, perhaps, two million light years and more away. This is a large spiral galaxy, estimated to be twice the size of our own; and it is a frequently studied galaxy because, despite its great distance, it is still close enough for astronomers to conduct very detailed studies at various wavelengths. Andromeda appears to be a star system very much like our own; it appears also to contain a wide variety of stars in various stages of evolution, and some very diverse nebulae. Andromeda quite possibly contains 100 billion stars and the laws of probability dictate that some of these stars must contain their own planetary systems. Currently, dozens of sources of X-rays have been detected within Andromeda, some coming from the compact inner bulge, some from globular clusters, and others from sources scattered about the galaxy.

The active galaxies are another matter all together. The X-ray and other radiation being emitted from these systems can be traced to relatively small and central locations. Astrophysicists suspect that the source of the strongest emissions from the majority of active galaxies is almost always a black hole. (They also suspect that a black hole is in existence at the center of our own galaxy, though no x-radiation has yet been detected from the center of the Galaxy). Some theorists explain that there must be a black hole at the center of all galaxies, normal or active, but in the normal galaxies the black holes are relatively small or the galactic nucleus itself is so barren that the black holes continually go unfed, so to speak. Since in these cases there is little or no debris to be sucked into the hole, there is insufficient friction being produced to create the temperatures needed to generate x-radiation that can be detected from Earth.

There are a number of types of active galaxies, besides radio galaxies, including, Seyfert and N galaxies. Of the two, the Seyferts are the stronger sources of X-rays, as well as ultraviolet and infrared rays. Seyferts may be classified as radio galaxies except that their radio emissions are not usually very strong and are sometimes almost undetectable. N galaxies are not always associated with either radio of X-ray emissions but are, rather, distinguished from other galaxies by their exceptionally bright nuclei.

## CLUSTERS OF GALAXIES

Millions of light years away are specks of light that represent not simply a star group, or perhaps one or two galaxies, but groups of galaxies that might consist of thousands of members the size of any star system in the Local Group. Galaxies generally appear in clusters throughout the universe. The Milky Way Galaxy and other members of the Local Group are part of a giant system of galaxies called the Local Supercluster. It may contain a hundred thousand or more galaxies, each of these galaxies containing 10 billion to 100 billion stars.

The galaxies contained in clusters like the Local Cluster are all bound to each other by gravitational interaction. They are moving out further into the universe by some as yet undiscovered mathematical relationship that keeps them sectioned off from other star systems of groups or star systems. The paradox is that redshift surveys seem to indicate that the gravitational interactions of these galactic clusters are not strong enough to keep each member galaxy from expanding. So, the Milky Way Galaxy, though tied in some cosmic bond with thousands of other galaxies in the Local Supercluster, can continue its expansion phase for at least another few billion years.

Actually, the Local Group is part of a giant, irregular cluster of galaxies contained within the Local Supercluster. This irregular cluster is the Virgo cluster, found within the constellation Virgo. It probably contains three or four thousand galaxies, although less than three thousand have been accounted for so far. Most of the galactic population consists of spiral galaxies, with the next most significant group being the ellipticals. Virgo is a strong X-ray source, as are several other clusters, particularly those which contain supergiant ellipticals.

X-ray studies of galactic clusters indicate that they are extremely different star configurations with varying degrees of luminosity and density. Luminosity, density, and configuration data give clue to the relative ages of these star systems, as well as clues to how they form. An irregular cluster emitting little x-radiation and having low central data may be categorized as an *carly system*. Those that are relatively symmetrical and have limited concentrations of x-radiation are called *intermediate systems*. And those that emit strong x-radiation, have central density and are highly symmetrical are called *evolved systems*.[14] Understanding the evolutionary phases associated with galactic clusters help astrophysicists

Fig. 9-6. VV 285 is a peculiar ring galaxy with a very violet Seyfert nucleus. Seyferts are like radio galaxies in many respects except that their radio emissions are not usually very strong (Courtesy of National Optical Astronomy Observatories).

determine the order in which galaxies and the clusters have formed—that is, it helps determine whether the galaxies evolved first and then were gravitationally forced into clusters, or the clusters existed first as some great nebula grid from which the individual clusters, or groups of clusters, were born. Other important astronomical questions, such as that of the missing mass, may also be answered once astronomers understand the way in which the clusters formed.

## BL LACS

Earlier, you were introduced to the strange and distant objects that appear to be no more than tiny dots of bluish light in the sky, yet tiny dots that give off the kind of energy astronomers expect to detect only from a cluster of galaxies. Photographically, these dots appear to be no more than stars and they might easily be mistaken for stars except for their very marked redshifts, which indicate they are moving very rapidly away from Earth. And though they are intensely bright, they emit mainly in the infrared, yet they are strong sources of X-ray emissions.

Well, there exist objects lying well beyond our galaxy which have many of the same characteristics as quasars, but spectral studies reveal the absence of both emission and absorption lines. (Emission lines are patterns in a spectrum caused by the transmission of electromagnetic radiation; and absorption lines are patterns produced in a spectrum when electromagnetic radiation is absorbed by matter.) These objects are called BL Lacs, and like quasars, they are very strong radio sources, though they do indeed emit mainly in the infrared.

BL Lacs are stormy systems. That is, they increase unexpectedly in luminosity by a factor of 100 or more and then suddenly dim again. It has been impossible for astronomers to determine their distance but preliminary investigations, based on measuring the redshifts associated with what might be surrounding clouds or galaxies in some instances, put BL Lacs as far out as the quasars. It appears the main distinction between BL Lacs and quasars, aside from emission and absorption data, is that the BL Lacs are associated in some way with elliptical galaxies and the quasars with spiral galaxies.[15]

X-ray studies of the BL Lac objects help astronomers learn more about the X-ray background detected throughout the universe, and possibly a bit more about quasars, and BL Lacs themselves, for BL Lacs are as much a mystery at this time as quasars.

## COSMIC X-RAY BACKGROUND RADIATION

It would be a shortcoming to close this chapter without additional discussion of the strong background presence of x-radiation discovered by Riccardo Giacconi during those first experiments with X-rays in astronomy. As subsequent observations have verified, in every direction in which instruments are pointed, X-rays are detected in nearly equal strength—just as it has been observed for microwave, infrared and gamma rays. And like microwave radiation phenomena, the background x-radiation is not coming from any celestial source but seems, rather, to be a product of happenings related to the very early evolutionary stages of the universe. Many astronomers are of the opinion that this background radiation is ''carrying data'' which, when finally decoded, will tell them much about the large scale structure of the universe.

Whereas microwave background radiation is believed to be associated with the very beginning of the universe, background x-radiation probably does not have its origins so early in cosmic history. Just exactly when background x-radiation was produced is, as anything else associated with cosmological ideas, subject to wild speculation. Astronomers believe the background radiation has its origins in a time when galaxies or clusters began to form, and, as it stands now, astronomers cannot really be sure of how galaxies formed or in what cosmic period their formation actually began.

One of the problems related to this background x-radiation is balanced on the question of whether or not the universe actually began with a Big Bang. The Big Bang theory has now just about completely replaced the steady state theory, but there are still some serious problems with the theory, not the least of which is the question: What was it that existed in the first place to allow the Big Bang to occur? After all, any explosion can only take place if there is something existing to explode. And expansion and evolution can only take place if there is something that can be extended and changed. To what did the primal substance belong, and of what did it consist? These are necessary questions, as everyone knows that nothing cannot produce something.

There are other conundrums associated with the Big Bang, having to do with the small amount of antimatter in the universe, the amount of time that must be allotted for the formation of galaxies, and the isotropic nature of the universe.[16] If indeed there was a Big Bang many billions of years ago, there must be in existence extensive populations of antimatter as well as populations of matter. As far as astrophysicists can determine, the Milky Way Galaxy and all other galaxies, contain no antimatter. If galaxies were indeed formed from condensing clouds, as cosmologists seem to expect, then it would have taken a lot more time for galactic clusters or even galaxies to form, given a universe, say, some 10 billion years of age. And why were Penzias and Wilson able to detect the microwave radiation that seemed to be coming from all areas of the universe, especially when the Big Bang model allows for ''cosmic freezings'' (periodical drops in temperature that result in extreme environmental changes). These cosmic freezings would greatly alter the radiation from different parts of the universe.

There are theoretical solutions to these problems that challenge the Big Bang theory. To the question of the antimatter population, one answer is that there must have been a lot more matter than antimatter created during the first second after the Big Bang, and another is that in the early stages of cosmic evolution matter and antimatter were thrown together in some numerical imbalance that matter, which was in abundance, survived. Somewhere out in space, there must be galaxies of antimatter.

In answer to the question of the time period required for star systems to form, there is the answer that developmental defects occurred shortly after the Big Bang, allowing for the existence of enough mass early enough in the universe's evolution to allow the formation of galaxies.

To the answer of the isotropic nature of the universe, one answer is that the physics of the Big Bang allowed the universe to attain a state of thermodynamic equilibrium very early on, and that it has remained in that state since.

It is important to stress that while the Big Bang theory still enjoys wide support, it is by no means a flawless theory and it continues to be subject to serious challenge. Whether or not there was a Big Bang is only one of the many problems that must be faced in determining when the background x-radiation was produced. The second has to do with the true age of the universe, which seems to change with each passing decade, and which may now be anywhere from 10 billion to 100+ billion according to scientific source or only thousands of years according to religious historians. Astronomers will probably never learn the true age of the universe or how it actually came into being.

Still another problem related to determining how the background x-radiation was produced is the way in which galaxies have formed. There are a lot of questions to be answered on this subject, but within the framework of current Big Bang models, it is estimated that about 10 billion years after Point Alpha (or Time Zero), clusters of galaxies began to form. It was during this time that the background x-radiation must have been produced. As Tucker and Giacconi point out, however, it appears to be highly improbable that the gas clouds from which the galaxies were formed could have reached the 400 million degrees required to produce the background radiation; and it is also improbable that the galactic nebulae could have been heated to this temperature by quasars or exploding galaxies.[18]

Still, given all the problems associated with Big Bang theories and the evolution of galaxies, astronomers hope that X-ray studies will help answer the question of whether the universe is open or closed. Current theory has it that the universe began with a fantastic explosion which threw out energy and matter with such force that it is still rushing out toward the unknown. Much of this matter has clumped into galactic clusters, stars, planets, and other major celestial bodies, but the great expansion that began with the Big Bang is believed to still be continuing. The galaxies, including our own, are still expanding, and still receding from each other. The great question remains: how long will the expansion continue before gravitational forces slow it down and a period of contraction sets in, this period of contraction ultimately resulting in a cataclysmic death of the universe as we know it, resulting in a return of the universe to what it must have been at the moment of the Big Bang? Or will the expansion continue indefinitely?

One of the ways in which astronomers and astrophysicists hope to answer this great question of expansion or contraction is to determine the relationship between gravity and mass in the universe. This subject has already been discussed briefly, but now let's take it a bit further. If there is enough material existing in the form of galaxies, then eventually gravitational interactions must necessarily halt the expansion. So all astronomers have to do is measure the mass and gravitation, and the rest is just a few complicated equations. The problem is astronomers do not have all the values to plug into these equations, for, as we have learned, much of the mass in the universe cannot be found. Here is where X-ray astronomy will play its most important role. In the past, astronomers have interpreted the background x-radiation to mean that the mass exists in the form of an intensely hot cosmic gas pervading throughout the universe—until recent satellite based observations. Rather than this background radiation being some intensely hot and all-pervading field of gas, results from NASA's High Energy Astronomy Observatory have revealed that the X-ray background is more likely the result of emissions from galaxies, galactic clusters and quasars. This means that the universe is probably open.

# Chapter 10

# Gamma Ray Astronomy

GAMMA RAYS ARE ELECTROMAGNETIC EMISSIONS of the shortest wavelengths and the highest frequencies. They may be produced by transitions between quantum energy levels in atomic nuclei, through the collisions of particles and antiparticles, during the decay of elementary particles, and through the acceleration of charged particles.[1] The gamma ray region overlaps the X-ray region of the spectrum and, generally, gamma rays can be divided into five levels:[2]

- Lower energy gamma rays ($10^5$ electronvolts)
- Medium energy gamma rays ($10^7$ electronvolts)
- High energy gamma rays ($10^9$ electronvolts)
- Very high energy gamma rays ($10^{11}$ electronvolts)
- Ultra high energy gamma rays ($10^{14}$ electronvolts)

## GAMMA RAY ASTRONOMY

Gamma ray astronomers are faced with the same problems of atmospheric attenuation that X-ray, ultraviolet, and infrared astronomers almost always face: their observations must be carried out from above the atmosphere. They also face the same difficulties with background radiation that others working in most areas of invisible astronomy must face, and they must, therefore, make extreme efforts to assure that they can distinguish gamma ray sources that may be coming from astronomical objects from the gamma rays that appear to be a part of background gamma radiation.

The first detection of gamma rays from an astronomical source occurred in the early 1960s during X-ray surveys of the Moon, but according to Hiller, the very beginnings of gamma ray astronomy must be traced to a research paper published in 1958—not because the paper detailed gamma ray astronomy, but mainly because it served as an incentive to begin looking for gamma ray sources in the sky. Later, in 1968, the first cosmic X-rays were detected by an Orbiting Solar Observatory (OSO) equipped with a scintillating counter. The amount of data accumulated throughout the 60's was scant and without much value. There was enough accomplished to whet continued interest and, in 1972 and 1975, satellites were launched specifically for the purpose of detecting gamma radiation from astronomical sources. These satellites were NASA's SAS-2 and Europe's COS B.

SAS 2 (Small Astronomical Satellite) followed in the path of the Uhuru satellite (SAS 1), launched specifically for X-ray research. SAS 2 was equipped for gamma ray detection, and was able to compile an impressive quantity of gamma ray data for analysis. COS B was launched by the European Space Agency in 1975 specifically to gather data on astronomical sources. SAS 2 recognized interstellar gas as a gamma ray source and pulsars as another. The highly successful COS B satellite was able to identify many specific quasars and molecular clouds as gamma ray sources.

Since those first detections of gamma rays in the 1960s, a multitude of sources have been detected, including the center of our own galaxy, and other galaxies as well. Additionally, the diffuse and isotropic flux that serves as a steady background of gamma radiation has been verified. Whether or not this cosmic flux has cosmological implications—as microwave background radiation is expected to have—is still to be determined. However, astronomers generally feel that the gamma ray background has little to do with cosmic evolution but is rather generated from very distant but very active galaxies, or else is resulting from subatomic interactions involving high energy protons.

## GAMMA RAY TELESCOPES

Atmospheric interference requires that gamma ray telescopic surveys be conducted from artificial satellites. However, in some cases, balloons are used. At this writing, NASA is planning to launch a satellite, the Gamma Ray Observatory, solely for the purpose of advanced and detailed studies of gamma ray sources. By the time this book makes it into bookstores, sometime in 1987 or 1988, the satellite may have already been launched.

Gamma rays cannot be observed by any optical means for, after all, they are invisible forms of radiation; therefore, the telescopes used for collecting gamma rays must be sophisticated electromechanical devices designed to translate detection into electronic data. Actually, all astronomical telescopes serve this same function; even optical telescopes used in the major observatories translate visual signals into electronic data for display in one form or another. This is why astronomers working in the visible ranges or the electromagnetic spectrum rarely get involved in direct observation; they more often can be found looking at a computer printout other than into an eyepiece.

Gamma ray detection is dependent upon the sensing and measuring of electrons ejected whenever gamma rays act and react with a given environment. Thus, telescopes must be specifically designed for the regions of the spectrum which are part of the subject for study. Certain types of design are necessary for low energy gamma ray detection and measurement, and others are required for high energy gamma rays. In either case, the telescopes used must be able to detect the direction of the gamma rays as well as their presence. Basically, gamma ray telescopes consist of detectors positioned behind a collimator. The purpose of the detector is to measure the photon energy being received. Usually these are scintillation detectors.

There are often serious problems in measuring and interpreting data collected by gamma ray telescopes. The problems occur for a number of reasons, from the inherent problems in designing the detecting systems to the presence of background radiation. It is not always easy to determine the source of gamma radiation because of the way in which low energy electrons induce multiple scattering in the detector; often the probable source of the gamma ray comes across as a blurring spot, and the exact point of origin can never be determined with 100 percent accuracy. It is impossible to measure gamma radiation with any great accuracy because the rays are weakened during scattering and the total energy present cannot be accurately sensed in short-interval studies. Gamma rays are generally weak anyway, and blend all too well into the prevailing background gamma radiation. Long exposure times (weeks) are often required to obtain sufficient spectral data.

For the detection and measurement of high energy gamma rays, telescopes are designed with small spark chambers that respond to the presence of electrons. Sensing devices included for use with the spark chambers include photomultiplier tubes. There is an anticoincidence system included as part of the telescope to ensure that the spark chamber is only activated by gamma rays. Spark chambers cannot be used in the detection of low-energy gamma rays because here the chamber's material inhibits penetration by the radiation and the detectors are not triggered. The spark chamber is a miniaturized version of those types of chambers used in particle accelerators; it consists of a series of metal plates arranged parallel within a gas-filled chamber. The gas is usually some mixture of neon and argon. Every plate is connected to ground but every other plate is designed so that with each passage of a charged particle, a relatively high voltage is received. This eventually results in sparking between the plates lying in the path of the incoming particles, and, so, the rays are detected.[4]

## SOLAR GAMMA RAYS

The Sun is not a major source of gamma rays. Thus, while X-ray studies have revealed a great deal about the Sun, gamma ray studies have so far been unable to add much to what we know about our nearest star. Most of the radiation which comes from the Sun originates somewhere in the photosphere. As temperatures here are only about 6,000 degrees, gamma ray emissions are very, very slight. In the case of solar flares, astronomers have found a source for radiation at a number of ranges in the spectrum.

X-ray emissions from flares are produced during the first increase in flare brightness. This means that the X-rays are produced at almost exactly the same time as microwaves. Microwave emissions, which peak somewhere in the neighborhood of 3000 megahertz, are synchrotronic in nature and are produced from the same electrons that create X-rays.

Flare phenomenon has come to be described in terms of five levels of intensity, or groups. The first group consists of the microwave bursts, which occur during the more explosive phases of flare evolution. Type II bursts are extremely intense and are characterized by a diminishing frequency range over a period of less than five minutes; this type of burst is probably caused by energetic particles crossing through the area of a flare. Type III bursts are much more frequent than Type II, although they are caused by the same particle movement; they show high frequency drifts which appear to be ties directly to coronal activity. Type IV bursts generally occur after Type II bursts as extensive wide band emissions. Type V bursts are generally after-effects of Type II bursts.

Solar flares have also been observed to occasionally be the source of cosmic ray particles. Cosmic ray particles are extremely high energy particles that race into Earth's atmosphere at almost the speed of light. Now, solar flares are not only divided into the five types typical of their spectral character and polarization tendency, but also into five levels based on the area which they take up. Only flares covering areas of 600 millionths of hemisphere are usually the source of these cosmic ray particles. Lower energy cosmic ray particles that originate from the Sun appear to be emitted from the solar photosphere and corona as well as from solar flares.

## GALACTIC SOURCES

Just about every light which is visible in the sky belongs to the Galaxy. (The only exceptions are a few small milky sources of light which belong to some nearby galaxies.) There is a faint field of light, mostly provided by stars which form a belt around the celestial sphere, separating the sphere into two distinct hemispheres. This stellar belt is the Milky Way. It is the main source of information about our Galaxy, so the entire Milky Way is a highly important "laboratory" for astronomers.

Our Galaxy is a spiral Galaxy possibly containing hundreds of billions of stars. It takes its name from that band of stars running across the celestial sphere and is, therefore, called the Milky Way System, or sometimes the Milky Way Galaxy.

This is a relatively flat Galaxy, containing most of its stellar population in a narrow disk having a bulging nucleus at its center. This nucleus lies in the con-

Table 10-1. Gamma Ray Astronomy.

| System | Major Sources |
| --- | --- |
| Galactic | Milky Way, cosmic ray interactions |
| Extragalactic | Crab and Nebula pulsars, gamma ray bursts |

stellation Sagittarius. Around the disk and its nucleus is a circle of stars and globular clusters which form what appears to be very much like a halo, but the celestial bodies making up this halo are much more loosely packed—far more scattered—than the objects which appear to over-populate the nucleus. Spiral arms extend from the region of the nucleus within, or very close to, the central plane.

The galactic disk is possibly 20 kiloparsecs in radius and possibly 4 kiloparsecs thick at the center. The bulge which appears at the center of the nucleus is estimated to be 300 parsecs thick and over 4500 parsecs in length. As far as the dimensions of the halo, astronomers are not particularly interested in guessing, but its radius is far greater than that of the disc.

Different areas of the Galaxy are populated by different kinds of stars. In and around the nucleus Population II stars can be found. Population II stars are the ancient reds that can also be found throughout elliptical and lenticular galaxies. These are high velocity stars and long period variables. There may also be subdwarfs, RR Lyrae stars, and Type II Cepheids. These Type II Cepheids are the older yellow giants or supergiants with pulsating periods of up to 70 days.

In the spiral arms and in and around the galactic disk population I stars can be found. These are very young stars with a strong metal content and very high luminosity. Older population I stars like the Sun, A stars, and Me dwarfs, are found within the halo as well as being scattered elsewhere about the Galaxy.

Earth is located, along with the rest of the solar system, between 8 and 15 parsecs from the central plane of the Galaxy, just inside one of the spiral arms.

The Galaxy rotates about a central axis—that is, the entire Galaxy is in continual motion about this axis, although all areas do not exhibit equal rotational speed. The disc moves with greater velocity than the halo. Most of the motion occurs in a circular scheme and appears to run parallel to the galactic plane, but the motion of specific objects may vary considerably from those more coincident with the general galactic movement. (One is reminded of Plato's cosmological scheme when considering the way in which the Galaxy moves.) At the same time that the Galaxy moves about this central axis, it is moving through space—or, to put it in terms of Big Bang cosmology, it is moving outward *with* space.

Those who are new to the subject of astronomy guess that galaxies are convenient groupings of stars that over time have come to be gravitationally independent, so that in the chronology of things, stars came first, and those that have formed a system have come to be classified in terms of galaxies. But this is hardly true. The galaxies existed first (as great clouds of gas and dust) and gave birth to the stars.

Take, for example, the Milky Way System. It began as a spherically shaped mass of gas maybe countless parsecs in radius. As time passed, this great cloud of gas began to contract, resulting in the formation of stars, the stellar productions continuing and increasing as the galactic cloud mass became more and more dense. By the time it had completed its collapse, the cloud had come to form the disc-like figure with its bulging nucleus that is typically a component of spiral galaxies. Though it appears to have completed its evolutionary trip from cloud to star system, the Galaxy is still a fertile field. Stars are continually being produced, especially in and around the spiral arms where there are strong concentrations of interstellar matter. This continuing production of stars is apparently a ceaseless phenomenon, because the material from which new stars are formed is continually produced through supernova explosions and other cosmic occurrences.

However, starbirth will not always be at a predictable rate. Right now, stars are not being produced quite as often as they were during the contraction of the great galactic cloud, and astronomers expect they probably will not be produced quite so often in the future unless galactic physics changes. There is little reason to assume that with each passing millenium, star birth will always be less than it was in the past. The extraordinary system we call Nature has a way of keeping things in balance.

Galaxies are not made up simply of stars. They are also populated by great gas and dust clouds and all kinds of objects. For instance, the Milky Way System includes planets, comets, black holes, neutron stars, and many soon-to-be-discovered objects.

From a general point of view, the Galaxy, therefore, may be considered a celestial object consisting of a nucleus, a disk, spiral arms and a halo. Extending far beyond the halo is an atmospheric field quite aptly named the Galaxy's corona. Bearing these simple facts in mind, let's consider the sources of the gamma ray emissions that appear to come from different directions in the Galaxy.

At gamma ray wavelengths, the Galaxy is a very different object than it is at other wavelengths, particularly the optical. At visible wavelengths, the Galaxy is a system of heavy concentrations of bright and dim lights representing stars and dust clouds. At radio and X-ray wavelengths, great gas clouds also become visible. At gamma ray wavelengths, the Galaxy appears as a very barren place, barren of stars in most areas except along the galactic plane. There are notable exceptions, including the constellations Crab and Vela. (The reasons for these exceptions will be explained shortly.)

The Sas-2 and COS B satellite experiments in the 1970s confirmed the Galactic disk as one of the major sources of gamma radiation. Studies of the energy ranges in which gamma rays were detected gave indication of the possible physical sources of the radiation. These have been determined to be the evolution of neutral pions, the rapid deceleration of electrons, and the interaction between photons and some charged particles.

Pions are elementary particles that have zero spin and are neutral, positively charged, or negatively charged. The charged particles are particles and antiparticles, but the neutral pion is its own antiparticle. The pion decays very quickly but it can travel up to 10 meters before doing so. Charged pions decay into muons and electrons, but the neutral pions almost always become gamma ray photons with periods of $8.4 \times 10^{-15}$ seconds.

So much for the evolution of pions into gamma rays. Now, we will consider the next source of gamma ray emissions, the deceleration of electrons. In this case, what happens is the production of a process called *bremsstrahlung*. Bremsstrahlung always results in the production of electromagnetic radiation.

The third source of gamma ray emissions at the Galactic disk is the interaction between photons and charged particles. This process is referred to as the Compton effect and it comes about as the result of a portion of the photon energy being absorbed by a charged particle. This means the photon cannot maintain (loses) its frequency of radiation at the same time that the particle gains as an energy field. When this process occurs in starlight, gamma radiation is produced.

Heller points out that gamma ray production as the result of bremsstrahlung is most frequent at the lower energies and gamma ray production as the result of pion decay is most frequent at the higher energies.[5] Both of these production methods result in narrow distributions in latitude because they occur in those regions where there are fertile fields of interstellar gas. Gamma ray occurrence as the result of the Compton effect (or Compton scattering) is on a much wider scale because in this method of production the only limitation is the diffusion of cosmic ray electrons.

Besides the diffuse radiation that has been collected from the Galactic disk, there are also discrete sources which include pulsars (supernova remnants), molecular clouds, the Galactic center and some mysterious sources.

Pulsars, as you have already learned, are very strong radio sources and, in fact, were first discovered by radio astronomers. There are also X-ray pulsars having pulsating periods of a few seconds to a few minutes; and pulsars which are gamma ray sources. The very first pulsars to be detected were discovered only a couple of decades ago, in 1968, by researchers at the Mullard Radio Astronomy Observatory, which is a part of Cambridge University. Hundreds of pulsars have since been detected. As mentioned in the chapter on radio astronomy, pulsars are believed to be rotating neutron stars. Estimates of their population range from a few hundred thousand to more than a million, and they are believed to occur at the rate of about six to 10 every one hundred years or so, which is somewhat more than the production rate of supernovae. This hints that there may be other ways in which pulsars are formed besides from the catastropic explosions associated with the gravitational collapse of massive stars.

Now, as a neutron star is expected to be what remains of a supernova explosion, pulsars, supernova remnants and neutron stars are one and the same thing. It is probably only the newly formed neutron stars that actually behave as pulsars, because these represent that stage of the supernova remnant when it is still reeling from the explosion which produced it and, therefore, still has an unusually rapid spin and an incredibly strong magnetic field. As most supernovae emit that nonthermal type of radiation known as synchroton radiation, they are necessarily producers of extremely high energy electrons and quite conceivably the source of discrete gamma radiation.

The only localized sources that can definitely be identified are those of the Crab and Vela pulsars. The Crab pulsar is a remnant of the same supernova explosion that produced the Crab nebula. The Vela pulsar is the brightest gamma ray source in the sky, but it is not a new pulsar and its rate of rotation is decreasing at a steady and probably swift rate. The very specific way in which gamma ray emissions from pulsars like Crab and Vela are produced, however, is still under investigation. The majority of the theoretical models assume a neutrons star's magnetic field to be the source of the gamma rays and that the pulsating signals are caused by the lack of coincidence between magnetic and rotational axis.[6]

In the case of molecular clouds, there is sufficient evidence to label these also as gamma ray sources, though only a few of the clouds have been sufficiently examined, among them a giant cloud in Orion. These moleculars are predominantly hydrogen clouds ($H_2$) with about a 1 percent dust content. Their average linear dimension is roughly 40 parsecs and they have the equivalent of roughly 500 thousand solar masses.[7] Their temperatures may reach to 20 degrees K but are rarely more than that; these are very cold clouds. The clouds generally contain more than one core region, these being sources for infrared, ultraviolet, X-rays, and cosmic rays.

Molecular clouds represent a great percent of the interstellar gas existing in the Galaxy. There may be more than 5000 of these clouds in existence, their total mass about 2 billion times that of the Sun. The Orion molecular cloud, of which the Orion Nebula is only a part, is believed to be about 200 solar masses. Another cloud, the Sagitarius B2 cloud located near the galactic center, is believed to be 200 thousand solar masses. The gamma ray emissions that are generated from regions within these clouds probably result from the way in which cosmic rays act and react with other constituents of the clouds.

The next source of gamma ray emissions within the Galaxy that appears on our list is the Galaxy's innermost region, the galactic center. It has been observed to be a very strong source of electron-positron annihilation radiation. This type of radiation results from the destruction of the above particle and antiparticle and the resulting production of the gamma ray photon. But, as you have learned in previous chapters, the galactic center is also a source of radio, infrared, and X-ray emissions. To this writing, most of the knowledge which has been accumulated about the galactic center is indeed owed to radio, infrared, and X-ray astronomy, but gamma ray astronomy may play a more significant part in research in the future.

This brings us now to the last remaining sources of gamma ray emissions within the Galaxy. The remaining emissions seem to come from mysterious sources which refuse to be identified. Astronomers guess that these unidentified sources may actually be interstellar regions where a number of phenomena interact to produce gamma radiation. Among the possible cosmic activities that may result in gamma ray emissions are the interactions of clouds and supernova material, and, quite possibly, the nebulae shielding very young stars.

## EXTRAGALACTIC SOURCES

When in 1988 NASA launches the satellite specifically designed to function as the U.S. Gamma Ray Observatory (GRO), a new phase in gamma ray astronomy will begin. The observatory will be equipped to detect gamma ray spectral features emitted by the nuclei of chemical elements recently synthesized during supernova explosions that have occurred as far away as the Virgo cluster of galaxies. In the Virgo cluster, a supernova explosion seems to occur at least once every year. This is a giant irregular cluster probably in the neighborhood of 15 to 20 megaparsecs away. There are more than 2000 galaxies in its system, yet the cluster itself is only one of those making up that 30 megaparsec cloud called the Local Supercluster, to which the Milky Way Galaxy and other members of the Local Group belong.

So, while a great many discoveries lie in the future, it is already expected that a significant number of galaxies will turn up to be strong gamma ray sources. This is because observations at radio wavelengths have already indicated the presence of high energy particles, and this means a good chance of gamma ray production. Additionally, it appears that these high energy sources make up relatively compact galactic areas which, because of their special characteristics, which include very powerful magnetic fields, are excellent candidates for the generation of gamma rays.[8]

The particular types of galaxies are Seyferts, radio galaxies and, of course, quasars.

The Seyferts, the radio galaxies, and the quasars all come under the heading of *active galaxies*, galaxies distinguished from other types because they emit relatively large amounts of radiation from highly compact regions at or near their centers. Seyferts are usually spiral galaxies producing electromagnetic radiation as a result of particle acceleration though their emissions are not produced by thermal means; that is, the particles producing the radiation are not interacting within the typically hot and very dense areas in which thermal radiation is produced. These Seyferts are optical, ultraviolet, infrared and X-ray sources, but they are predominantly infrared objects. That is, if they are observed at these wavelengths, they produce a more extensive image at the infrared. Seyferts are probably quasars of some sort, though they are less powerful sources of energy. As mentioned earlier, it is currently expected that Seyferts are not really a type of galaxy as much as they represent a stage in the evolution of a galaxy. All galaxies are expected to behave at one time or another like a Seyfert.

Seyfert galaxies can be classified into two main categories. In Type I Seyferts, there is a spectral resemblance to quasars and evidence of strong velocities. In Type II Seyferts, the spectral characteristics are very different from quasars and there is strong evidence of lower velocity. Type I's may be considered definite sources of gamma radiation and, indeed, the closest Seyfert, NGC 4151, has been determined to be a source of gamma radiation. In the case of NGC 4151, it is expected that the region of emission at its center is governed by a black hole. In most cases, the energy source within Seyferts is no more than a few light months across.

Like Seyferts, radio galaxies are active galaxies, and these types of astronomical objects were first introduced in Chapter 5. It is important to note that the term "radio galaxy" actually only refers to one single type of galaxy producing radio emissions. These types of galaxies are the select group of ellipticals that produce very intense radiation at the radio wavelengths. They are such powerful sources of radio emissions because of the field from which the radiation appears to come. In most of the galaxies which give off radio signals, the emissions are coming from highly limited areas within the galactic structure. In the case of radio galaxies, the emissions are coming from all parts of the ellipticals at the same time. Additionally, in most of these radio galaxies, the radiation is also coming in great strength from regions just beyond the galactic parameters.

It also appears that radio galaxies have two very distinct energy sources. These sources may lie at the extremes of the Galaxy. Each source seems to be connected by an energy field to the other. This characteristic of separate energy fields is one important way of distinguishing the radio galaxies from the "normal" galaxies like those in the Local Group, which also give off radio emissions. Some of these radio galaxies are giant ellipticals many times the size of the largest galaxies to be found in the Local Group.

The way in which radio galaxies produce their energy is still to be determined. The current consensus is that these radio signals are synchrotron emissions, but the gamma ray signals are Compton radiation.

This brings us to quasars, those very distant objects that give off far more light than all the galaxies in the Local Supercluster.

Quasars are mainly infrared sources, but they are also radio, X-ray, and gamma ray sources. Not all quasars are radio sources and those that do emit at radio wavelengths do so in a range of 10m to 3mm.

Most of the observed quasars are believed to be more than 10 billion years old, which means that they were among the first major astronomical objects to be formed in the universe. They may have come into existence around the same time that the galaxies began their transitions from great clouds of dust and gas to extensive star systems. As a matter of fact, quasars may actually be galaxies or the nuclei of galaxies.

Astronomers estimate that quasars are, perhaps, a million times more massive than the Sun. Exactly how their energy is produced is as much a mystery as is the production of energy in radio galaxies. There are many theories, however, attempting to explain the mechanism for the quasars' radiation, some attributing black holes as the powerhouse, others explaining that supernova explosions or stellar collisions are the reason.

## GAMMA RAY BURSTS

In addition to these galactic and extragalactic known and unknown sources, there is another kind of gamma ray phenomenon mystifying astronomers: the very short but very powerful bursts of cosmic rays that produce energy rated at up to a million electronvolts. The first news of their existence came in an announcement by the Air Force in 1973, although they were first detected by Vella satellites in 1967.

These bursts generally last for less than 10 seconds, although some have been time to continue for more than a minute. Because these emissions come in flashes, astronomers have been considering flare stars and neutron stars as their possible sources.

Flare stars are relatively cool red stars that give off bursts of energy resulting in great increases in the brightness of the source star. These flare stars are not only optical objects, they are also radio, and X-ray objects, and can quite conceivably also be gamma ray objects. The flashes are probably produced because the source stars are very young stars and therefore rotating rather rapidly. Such rapid rotation would naturally produce the type of bursts quite similar to these mysterious gamma ray flashes. But the optical, radio and X-ray flares so far observed from celestial objects are sometimes produced by the interaction of binaries, and the gamma rays bursts may also have similar origin.

In the case of the neutron star, bursts of radiation are also produced by rapid spin. These neutron stars are supernova remnants still spinning from the force of the explosion that converted them. The younger a neutron star, the greater its spin and its magnetic field.

Thus, neutron or flare stars are suspected of being possible sources of gamma ray bursts because they are already producing flashes of radiation at other wavelengths in periods similar to that recorded for gamma ray bursts.

Neutron and flare stars are possible suspects only if the astronomical objects which produce them lie within the Galaxy. As these sources still cannot be identified, it is possible that these gamma ray bursts are originating from beyond the Galaxy. Arguments against this possibility, though not conclusive, are strong. They center on the luminosity, size, and temperature that would be required for an extragalactic source to produce such powerful bursts.

Astronomers will consider Type II supernovae as possible extragalactic sources. These are the types of supernovae that appear only to occur in the arms of spiral galaxies like the Milky Way Galaxy and Andromeda Galaxy. Type II supernovae come about as the result of the destruction of a supergiant star having more than eight times the mass of the Sun. The supergiant reaches a point when it is no longer able to continue the nuclear processes that provide its core energy, and so it collapses into either a neutron star or a black hole, depending upon just how massive it actually is. Perhaps supergiants with 20 solar masses or more explode into black holes.

Whether the source of these gamma rays lie within or without the Galaxy, or these bursts are the result of some stellar phenomenon or interstellar event, continues to be one of the major mysteries in astronomy today.

# Notes

## Chapter 1

1. William L. Langer, *An Encyclopedia of World History* (5th Ed.), Houghton Mifflin, Boston, 1972, p. 36.
2. A. Pannekoek, *A History of Astronomy*, Interscience, 1961, p. 84.
3. Robert Burnhan, Jr., *Burnhan's Celestial Handbook*, Vol. 1, p. 389.
4. Ibid., p. 392
5. A.T. Olmstead, *History of the Persian Empire*, Univ. of Chicago, 1948, p. 478.
6. A conclusion drawn from the great esteem in which Plato and Aristotle held the Magi.
7. John Gribbin, *Genesis: The Origins of the Universe*, Delacorte, New York, 1981, p. 5.
8. Pannekoek, p. 106.
9. Aristotle, *Metaphysics*, Book 1, Chapter 3.
10. Ibid.
11. W.L. Reese, *Dictionary of Philosophy and Religion: Eastern and Western Thought*, Humanities Press, Atlantic Highlands (N.J.), 1980, p. 636.
12. Aristotle, Chapter 4.
13. Ibid., Chapter 3.
14. Plato, *The Republic*
15. Aristotle, *On the Heavens*, Book 2, Chapter 1.
16. Ibid., Chapter 4.
17. Ibid., Chapter 7.
18. Ibid.
19. Ibid., Chapter 8.
20. Ibid., Chapter 9.
21. Ibid., Book 3, Chapter 14.
22. Pannekoek, p. 125.
23. Ibid.
24. His full name was Abu Jafar abd-Allah-Mansur. It was he who conceived the idea of the great city of Baghdad and was responsible for the beginning of its construction during the eighth year of his reign.
25. Probably ¼ of the known stars are double stars.
26. Dieter B. Hermann, *The History of Astronomy from Herschel to Hertzsprung*, Cambridge U. Press, London, 1984, p. 22.

27. Pannekoek, p. 312.
28. Hermann, p. 75.
29. Ibid., p. 200.
30. Ibid., p. 126.

### Chapter 2

1. J. Bernard Cohen, *Isaac Newton's Papers & Letters On Natural Philosophy and Related Documents, 2nd ed.*, Harvard U. Press, Cambridge, Mass., 1958, p. 53.
2. Ibid.
3. George W. Stroke, "Light," in *McGraw-Hill Encyclopedia of Physics*, New York, pp. 552-53.
4. The Michelson-Morley experiments were an important contribution to the relativity theories of Albert Einstein.
5. Albert Einstein and Leopold Infeld, *The Evolution of Physics*, Simon & Schuster, New York, 1954, p. 274.
6. Ibid.
7. Ibid., p. 275.
8. Ibid., p. 278.
9. Cohen, p. 53.
10. Ibid.
11. Ibid., p. 54.
12. Ibid.
13. W.D. Niven, ed., *The Scientific Papers of James Clerk Maxwell*, Dover, New York, 1890, p. 136.
14. Ibid.
15. Bruce Billings, "Polarized Light," in *McGraw-Hill Encyclopedia of Physics,* New York, pp. 859-60.
16. Dudley Williams, "Particles," in *McGraw-Hill Encyclopedia of Physics*, New York, p. 776.
17. Ibid.
18. Ibid., p. 777.
19. Ibid.
20. George W. Stroke, "The Doppler Effect," in *McGraw-Hill Encyclopedia of Physics*, New York, p. 257.
21. John Gribbin, *In Search of the Big Bang*, Bantam, New York, 1986.
22. James S. Trefil, "Concentric Clues from Growth Rings Unlock the Past," *Smithsonian*, July 1985.
23. George E. Williams, "The Solar Cycle in Precambrian Time," *Scientific American*, August 1986, p. 88.
24. Charles R. Pellegrino, *TIME GATE, Hurtling Backward Through History*, TAB, Blue Ridge Summit, Pa., 1985.

### Chapter 3

1. Gerard De Vaucouleurs, *Discovery of the Universe: An Outline of the History of Astronomy from the Origin to 1956*, Maxmillan, New York, 1957, p. 146.
2. Michael Zeilik, *Astronomy: The Evolving Universe*, Harper & Rowe, New York, 1982, p. 90.
3. De Vaucouleurs, p. 146.
4. "Did an Asteroid Create the Everglades?," *Astronomy*, May, 1986, p. 69.
5. De Vaucouleurs, p. 153.
6. Ibid., p. 166.
7. Ibid., p. 167.

### Chapter 4

1. Marion Donnelly, *A Short History of Observatories*, University of Oregon Press, 1973, p. 5.
2. Ibid.
3. Richard Learner, *Astronomy Through the Telescope*, Van Nostrand Reinhold, New York, 1981, p. 19.
4. There had been a lot of criticism toward the project from the very beginning as well as some supportive press.
5. Learner, p. 101.
6. Donnelly, p. 102.
7. Marx Siegfried, Werner Pfau, *Observatories of the World*, Van Nostrand Reinhold, New York, 1981, p. 19.
8. Harold Zirin, "Sun," *McGraw-Hill Encylopedia of Astronomy*, New York, 1983, p. 39.

### Chapter 5

1. Hannes Alfven, Gustaf Arrhenius, *Evolution of the Solar System*, NASA, pp. 24-25.
2. For a detailed explanation of the evolution of the Earth and other planets, see Alfen and Arrhenius (above) and Clark R. Chapman, *The Inner Planets*, Charles Scribner, New York, 1978.

3. For a report on Giotto's achievements in tracking the comet, see Richard Berg's "Giotto Encounters Comet Halley," in *Astronomy*, June 1986, pp. 6-22.
4. Michael Rowan-Robinson, *The Cosmological Distance Ladder: distance and time in the universe*, W.H. Freeman, New York, p. 98.
5. Some of Hubble's more famous works were: *A general Study of Diffuse Galactic Nebulae* (1929),*Spiral Nebula as a Stella System* (1929); and *The Observational Approach to Cosmology* (1937).
6. Rowan-Robinson, p. 98.
7. Valerie Illingwood, ed., *The Facts on File Dictionary of Astronomy*, 2nd. edition, Lawrence Urdang, New York, 1985, p. 38.
8. Ibid., p. 239.
9. Ibid., p. 240.

### Chapter 6

1. For a very interesting and popular work on the Big Bang theory, see James L. Trefil's, *The Moment of Creation*, New York, Charles Scrib ner's, 1983.
2. As reported in *Astronomy*, August 1986, p. 64.
3. Charles J. Caes, *Cosmology: The Search for the Order of the Universe*, TAB, Blue Ridge Summit, Pa., 1986, p. 69.
4. As reported in *Astronomy*, May 1986, p. 69.
5. Charles J. Caes, *Beyond Time*, University of America Press, 1985, p. 12.
6. Ibid., p. 14.

### Chapter 7

1. Illingwood, p. 182.
2. A chemical radical consisting of an atom of hydrogen and an atom of oxygen; it has the typical characteristics of alcohol, glycol, hydroxide and oxygen oxide. OH was the first interstellar molecule to be detected in space; its discovery is owed to radio astronomy.

### Chapter 8

1. Ultraviolet light blackens photographic film and causes phosphorescent substances to give off special luminescence.
2. The way in which ultraviolet radiation interacts with a substance (ergosterol) produces vitamin D.
3. "Anniversaries and Events," *Air & Space*, July 1986, p. 20.
4. Skylab remained operable until about 1978 when it began to lose altitude. In July of 1979, while the whole world waited with great expectation and fear (that Skylab might crash in a populated area), Skylab crashed into the Indian Ocean.
5. A coronograph is used for observing the Sun's corona, which cannot normally be seen except during an eclipse because the corona is only a millionth as bright as the surface.
6. Illingwood, p. 20.
7. Ibid., p. 393.

### Chapter 9

1. Bruce R. Wheaton, *The Tiger and the Shark—Empirical Roots of Wave-Particle Dualism*, Cambridge U. Press, 1983.
2. William M. Sinton, "Telescope," in *McGraw-Hill Encyclopedia of Astronomy*, 1983, p. 399.
3. Illingwood, p. 297.
4. Ibid., p. 153.
5. Valerie Neal, *Renewing Solar Science, The Solar Maximum Repair Mission*, NASA (EP 206), p. 17.
6. Ibid., p. 8.
7. Ibid.
8. Wallace Tucker and Riccardo Giacconi, *The X-Ray Universe*, Harvard University Press, Cambridge, 1985, pp. 39-40.
9. Ibid., p. 42.
10. Ibid., pp. 44-6.
11. "High Energy Astronomy Observatory," NASA, p. 10.
12. Philip C. Peters, "Black Hole," *McGraw-Hill Encyclopedia of Astronomy*, 1983, p. 47.
13. Illingwood, p. 57 & p. 168.
14. Tucker and Giacconi, p. 176.
15. Illingwood, p. 45.
16. Charles J. Caes, Cosmology: *The Search for the Order of the Universe*, TAB, Blue Ridge Summit, Pa., 1986, pp. 72-74.

17. Ibid.
18. Tucker & Giacconi, p. 182.

## Chapter 10

1. Rodney Hiller, *Gamma Ray Astronomy*, Clarendon Press, Oxford, 1984, p. 3.
2. Illingwood, p. 151.
3. Hiller, p. 1.
4. See Hiller, Chapters 2-4, for a detailed discussion of gamma ray telescopes.
5. Ibid., p. 135.
6. Ibid.
7. Illingwood, p. 239.
8. Hiller, p. 178.

# Selected Bibliography

This is not meant to be a comprehensive list of all books and articles on the subject of astronomy and electromagnetic radiation; rather, it is a list of the books and articles that served as sources for the information presented in this text.


Alfven, Hannes; and Gustav Arrhenius, *Evolution of the Solar System. Washington: NASA, 1976.*

*Aristotle, "Physics," trans. R.P. Hardie and R.K. Gaye, in Aristotle*, ed. John Maynard Hutchins. Chicago: Encyclopedia Britannica Great Books of the Western World.

Aristotle, "Politics," in *Plato and Aristotle*, trans. Benjamin Jewett. New York Colonial Press, 1906.

Billings, Bruce, "Polarized Light," McGraw-Hill Encyclopedia of Physics, New York.

Brown, Hanbury, *Man and the Stars*. London: Oxford U. Press, 1978.

Burnham, Robert, Jr., *Burnham's Celestial Handbook*. Dover, 1978.

Caes, Charles J., *Beyond Time*, Washington: University Press of America, 1985.

——*Cosmology, The Search for the Order of the Universe*. Blue Ridge Summit, Pennsylvania: TAB BOOKS Inc., 1986.

Cohen, J. Bernard, *Revolutions in Science*, Cambridge: Belknap Press of Harvard U., 1985.

——*Isaac Newton's Papers and Letters on Natural Philosophy and Related Documents*, 2nd. ed. Cambridge: Harvard U. Press, 1978.

Davies, Paul, *God and the New Physics*, New York: Simon and Schuster, 1983.

De Vaucouleurs, Gerard, *Discovery of the Universe: An Outline of Astronomy from the Origin to 1956.* New York: Macmillan, 1957.

Donnelly, *A Short History of Observatories.* University of Oregon Press, 1973.

Dudley, William, "Particles," in *McGraw-Hill Encyclopedia of Physics*, New York.

Einstein, Albert; and Leopold Infeld, *The Evolution of Physics.* New York: Simon & Schuster, 1954.

Gribbin, John, *Genesis: The Origins of the Universe.* New York: Delacorte, 1981.

——*In Search of the Big Bang*, New York: Bantam, 1986.

Habing, Harm J. & G. Neugebauer, "The Infrared Sky," *Scientific American*, November 1984.

Hermann, Dieter B., *The History of Astronomy from Herschel to Hertzsprung*. London: Cambridge University Press, 1984.

Hiller, Rodney, *Gamma Ray Astronomy*. Oxford: Clarendon Press, 1984.

Illingwood, Valerie, ed., *The Facts on File Dictionary of Astronomy*, 2nd. ed. New York: Facts On File, 1985.

Kepler, Johannes, "Epitome of Copernican Astronomy," in *Copernicus, Kepler, Galileo*. Chicago: Encyclopedia Britannica Great Books of the Western World, 1948.

Langer, William L., *An Encyclopedia of World History (5th ed.)*. Boston: Houghton Mifflin, 1972.

Learner, Richard *Astronomy Through the Telescope*. New York: Von Nostrand Reinhold, 1981.

Mauldin, John H., *Particles in Nature, The Chronological Discovery of the New Physics*. Blue Ridge Summit, Pennsylvania: TAB BOOKS Inc., 1986.

Montmerle, T., "Gamma Ray Astronomy," McGraw-Hill Encyclopedia of Astronomy, 1983.

Neal, Valerie, *Renewing Solar Science, The Solar Maximum Repair Mission*, NASA (EP 206).

Niven, W.D., ed. *The Scientific Papers of John Clerk Maxwell*. New York: Dover, 1890.

Pannekoek, A., *A History of Astronomy*. New York: Interscience, 1961.

Pellegrino, Charles R., *Time Gate: Hurtling Backwards Through History*. Blue Ridge Summit, Pennsylvania, TAB BOOKS Inc., 1986.

Plato, "Timaeus," trans. Benjamin Jewett, in *Plato*. Chicago: Encyclopedia Britannica Great Books of the Western World, 1952.

Ptolemaeus, Claudius, "The Almagest," trans. R. Catesby Taliaferro, in *Ptolemy*. Chicago: Encyclopedia Great Books of the Western World, 1952.

Reese, W.L., *Dictionary of Philosophy and Religion: Eastern and Western Thought*. Atlantic Highlands, N.J.: Humanities Press, 1980.

Rowan-Robinson, Michael, *The Cosmological Distance Ladder: distance and time in the universe.*. New York: W.H. Freeman, 1985.

Siegfried, Marx, and Werner Pfau, *Observatories of the World*. New York: Von Nostrand, 1982.

Sinton, William M., "Telescope," *McGraw-Hill Encyclopedia of Astronomy*, 1983.

Stroke, George W., "Light," McGraw-Hill Encyclopedia of Physics.

Trefil, James S., "Concentric Clues from Growth Rings Unlock the Past," *Smithsonian*, July 1985.

Tucker, Wallace, and Riccardo Giacconi, *The X-Ray Universe*. Cambridge: Harvard University Press, 1985.

Williams, George E., "The Solar Cycle in Precambrian Time," *Scientific* American, August 1986.

Williams, S.A., "Amplitude," *McGraw-Hill Encyclopedia of Physics*

Zeilik, Michael, *Astronomy: The Evolving Universe*, New York: Harper & Row, 1982.

Zirin, Harold, "Sun," *McGraw-Hill Encyclopedia of Astronomy*, 1983.

# Index

# Index